S0-CKW-541

Environmental Geochemistry in Health and Disease

The Geological Society of America, Inc.
Memoir 123

Environmental Geochemistry in Health and Disease

American Association for
Advancement of Science Symposium
Dallas, Texas, December 1968

Edited by
HELEN L. CANNON
U.S. Geological Survey, Denver, Colorado
and
HOWARD C. HOPPS
The Armed Forces Institute of Pathology
Washington, D.C.

1971

Library of Congress Catalog Card Number 78-111440
I.S.B.N. 0-8137-1123-1

Published by
THE GEOLOGICAL SOCIETY OF AMERICA, INC.
Colorado Building, P.O. Box 1719
Boulder, Colorado 80302

Printed in the United States of America

The publication of this volume has been made possible
through generous contributions to the
Memoir Fund of The Geological Society of America, Inc.

Foreword

An interdisciplinary symposium on *Environmental Geochemistry in Relation to Human Health and Disease* was held on December 30, 1968, in Dallas, Texas, at the Annual Meeting of the American Association for the Advancement of Science. Participants included scientists from the fields of geochemistry, chemistry, soil science, geography, epidemiology, pathology, biochemistry, nutrition, and dentistry. The purpose of the symposium was to show that the chemistry of rocks, soils, plants, and water in a particular geographic environment may be causally related, either directly or indirectly, to the occurrence of animal and human diseases.

The data presented in this volume, based upon the Symposium described above, are representative of the state-of-knowledge in the field of **geochemistry** concerning the distribution of elements in various rock types of the substrata, the dispersal of these chemical constituents in soils and water during weathering and their absorption by plants. The variance in trace-element content from optimum levels of availability in different geographic areas is controlled by a combination of geologic and climatic conditions.

With respect to biomedical problems, the information presented here concentrates on medical ecology and the importance of geographic pathology in determining cause/effect relationships. Specifically, relationships of: zinc to body growth and wound healing; cadmium and hardness of water to heart disease; lead to multiple sclerosis; chromium to diabetes; molybdenum and strontium to dental caries; nickel, cadmium, chromium, and asbestos to cancer; and molybdenum to molybdenosis of cattle, are discussed in some detail.

The enthusiasm engendered during this Symposium led to organization of a Committee for continued multidisciplinary investigations of relationships between health and geochemical environment. More recently, a Subcommittee of the National Academy of Sciences has been established to further studies on Geochemical Environment in Relation to Health and Disease. We hope that those readers of this Symposium Volume who have pertinent information or knowledge about possible effects of trace element distribution on geographic patterns of disease will contribute to the state-of-the-art studies being made by the subcommittee.

The important role of The Geological Society of America in sponsoring this Symposium, and in developing the studies described herein, is gratefully acknowledged.

HELEN L. CANNON
HOWARD C. HOPPS

Contents

*Deceased, October 5, 1970

THE GEOLOGICAL SOCIETY OF AMERICA, INC.
MEMOIR 123, 1971

Geographic Pathology and the Medical Implications of Environmental Geochemistry

HOWARD C. HOPPS, M.D.
The Armed Forces Institute of Pathology
Washington, D.C.

ABSTRACT

A concept of disease ecology is presented, emphasizing the complex causality of disease. Four broad categories of causal factors are identified and discussed: "etiology," contributory causes, heredity, and environment. Specific illustrations are considered under each category, including hookworm infection, G6PD deficiency, skin color, gastric secretion, lead poisoning, schistosomiasis, blackfoot disease, the Balkan nephropathy, amyotrophic lateral sclerosis, and Parkinson's disease.

CONTENTS

Figure

INTRODUCTION

This year marks the centennial of Ernest H. Haeckel's (1834–1919) "Natural History of Creation," a treatise published in German in 1868 and, five years later, translated into English by the British zoologist, Edwin Ray Lancaster. Haeckel coined the term "Okologie" from two Greek words: *oikos,* meaning household, and *logos,* meaning discourse (in the sense of scholarly dissertation). He defined ecology (okologie) as "the body of knowledge concerning the economy of nature—the investigation of the total relations of the animal to its inorganic and organic environment."

This Symposium is about the ecology of disease, concentrating on those geochemical factors that may cause or affect disease. My objective is to describe and illustrate a point of view that regards disease in its broad context, considering its complex causality as well as the many factors that may alter its course and manifestations.

NATURE OF DISEASE

Perhaps the best way to get at the essential nature of disease is to look at health. Health reflects a relationship between the individual and his environment in which there has been effective adaptation, a continuous, dynamic, favorable adjustment to the antagonistic forces of his surroundings; the individual is **in balance with** his environment. Disease, on the other hand, is a state of imbalance—the consequence of maladjustment. Although we may speak of a disease entity, the term *entity* suggesting something static, disease is a **process,** one in which the body reacts to injury by a complex series of actions and counteractions that continue until the individual recovers or dies (*see* Fig. 1). Disease, then, is a **group** of effects that reflects the dynamic interaction of a group of causal factors with a group of host factors. These complex cause/effect relationships, indeed, the whole natural history of the process, is properly considered as the **ecology of disease.**

Much yet remains to be learned about disease ecology, but one thing has become clear; the old concept of disease, one cause/one effect, is no longer tenable—and an effective study of disease ecology must focus upon the multiple, complexly interrelated causes of disease. A schematic presentation of these multiple causal factors is shown in Figure 2, and much of my discussion will relate to this figure.

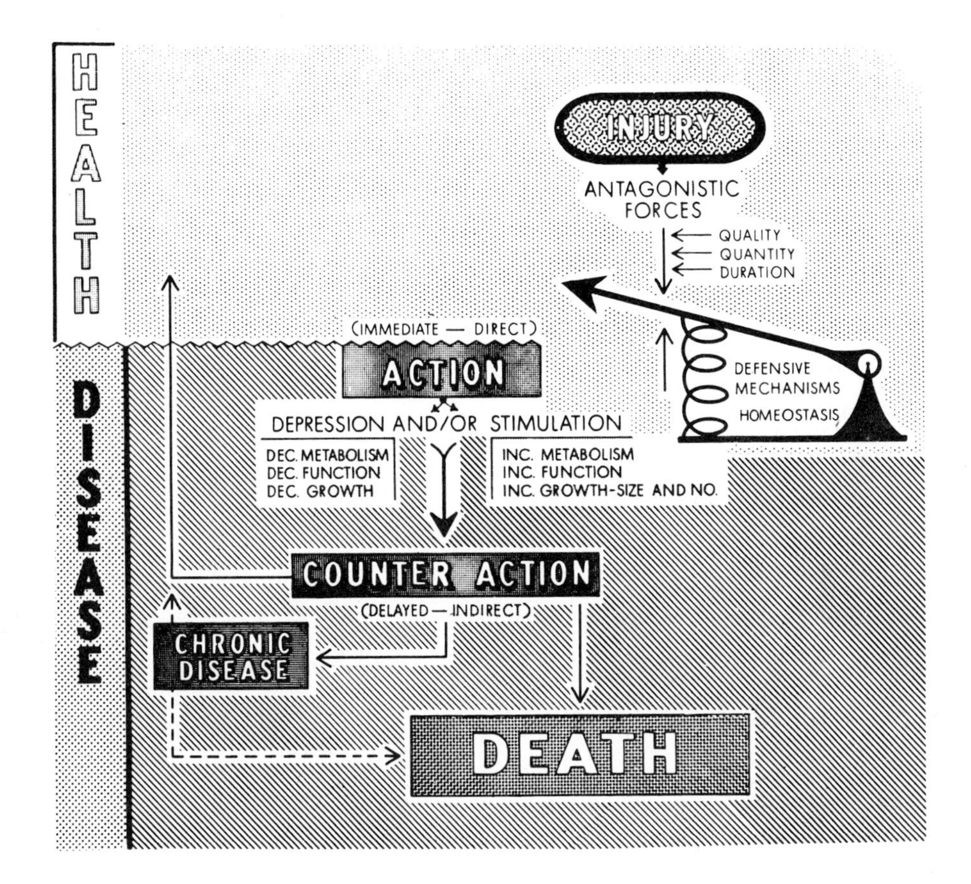

Figure 1. Disease may be viewed as a reaction to injury that produces an imbalance of forces. Following the stage of action comes counteraction. In contrast to the simple, immediate action, counteraction is almost always complex, indirect, and delayed. (*From* Hopps, H. C., 1964, Principles of Pathology, 2d ed.: New York, Appleton-Century-Crofts, reproduced by permission.)

CAUSES OF DISEASES

Let us begin analysis of Figure 2 by examining **"etiology"** and **contributory causes** in a general way. Etiology (Gr. *aitia,* cause + *-logy*) is most often a complex of interacting agents and factors, some of which reinforce, some of which antagonize each other. For example, initiating agents may work alone or only after a preparative agent has laid a necessary ground work. A promoting factor may be required if the agent is to reach a level of effectiveness, and a directive factor may be needed to "guide" the agent to those particular cells or tissues that have the required capacity for response. Frequently, these causal steps have to occur in a precise sequence and within sharply defined time limits to be effective. Furthermore, there may be rigorous requirements as to dose and portal of entry. Moreover, all of these causal agents and factors must act on a susceptible individual. Finally, a precipitating cause may be essential if the disease or condition is to become overt. This final step in the

Figure 2. The four major categories of causal factors are shown as though they were of equal value. Actually they rarely contribute equally to the complete cause of a disease. Each of the four broad categories includes subcategories, and sub-subcategories, as illustrated by the following:

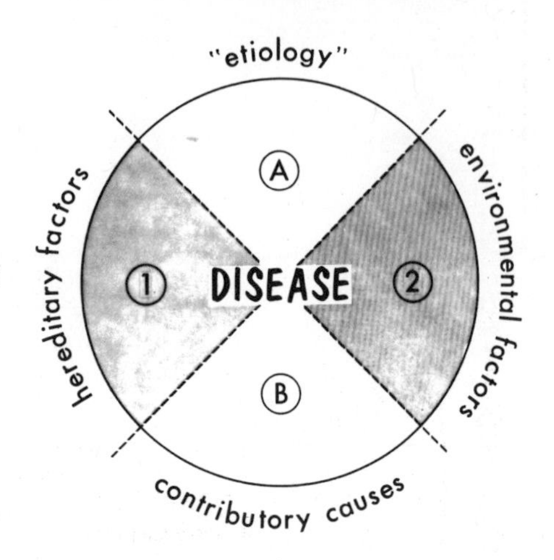

A. "ETIOLOGY"
e.g., specific agents:

physical
chemical
biological, including vectors and reservoirs

1. HEREDITARY FACTORS
e.g., specific genetic defects as manifested by:

G6PD deficiency
hemachromatosis
diabetes mellitus

e.g., genetically determined characteristics:

Hb structure
skin color
immunologic responsiveness
gastric secretion

B. CONTRIBUTORY CAUSES
e.g., exposure to the primary agents:

(with considerations of dose and portal of entry)

factors increasing susceptibility:
physical stress *(heat, cold, dehydration, low or high barometric pressure, exhaustion, trauma)*
mental stress
other diseases or conditions
(malnutrition, anemia, infection, cancer, drug therapy)

2. ENVIRONMENTAL FACTORS
e.g., geochemical
climatic
politico-economic
cultural

process is analogous to pulling the trigger of a gun that has been loaded with appropriate ammunition, cocked and pointed at a suitable target at a proper time. Often the immediate effects of a direct precipitating cause—the pulling of the trigger, as it were—are so impressive that we look upon this as **the** cause and restrict our thinking accordingly.

These are some of the reasons why it is so difficult to ferret out the primary, that is, initiating cause of many diseases, for example, the connection between too little zinc or too much nickel and a specific disease consequence: growth retardation or cancer.

Determining the Cause of a Disease. Efforts to connect primary causes with disease manifestations are further hampered because, more often than not, one is forced to deal with the end results of such complex interacting factors as have just been described. Trying to understand the etiology/pathogenesis from looking at the established disease is like trying to understand the anatomy of a murder from looking at the dead body. In disease investigation, just as in homicide investigation, one needs to know the circumstances of the final event, including the many events that lead up to it. Often, unfortunately, we do not know just how to recognize the critically important first stages. Sometimes it is like looking for a needle in a haystack, but without knowing that a needle is the object of the search. This is why our inquiry into the complete etiology of such conditions as atherosclerosis and cancer has been so difficult, and so slow.

Once the precise area that requires investigation is identified, we can turn to the laboratory and carry out experiments under controlled conditions, but here too we encounter problems of our own making. Eager to get dramatic, easily measurable results, we are likely to study the effect or effects of a single causal agent given in a single large dose. For example, in *in vivo* studies LD_{50} doses (lethal for 50 percent of the animals) of infectious agents or toxic substances are often used. Nature does not ordinarily operate this way. She is more likely to administer low-level doses repeatedly over a long, long time that act subtly in association with many other causal factors, some of which reinforce, others of which block the primary effects. Then, too, our experiments are usually designed to oversimplify the conditions, eliminating or minimizing all the factors except those under study. Of course this is the experimental method, and there is nothing wrong with such an approach (assuming that we recognize the major causal factors) so long as we maintain a proper perspective. The trouble comes when, in our enthusiasm and naiveté, we extrapolate directly from the simple to the complex. This fault is compounded by the tendency to consider a factor to be insignificant merely because we do not know how to control it or to measure its effects.

The first evidence of relationship between a disease and a causal factor, especially one connected to geochemical environment, may come from examination of distribution maps that show concurrence between the particular disease and the particular factor, and this is a powerful research method. It is easy, often tempting, to jump from mere association to causal relationship, but this large jump should be made cautiously. For example, McGlashan (1969, written commun.) has demonstrated that whooping cough is noticeably less frequent in the Zambian copper belt than elsewhere in that country. Rather than having any direct relation to metabolic activities of copper as a trace metal, however, he has shown that this anomalous prevalence of whooping cough is, in fact, due to immunization programs at the mine hospitals. Based on his extensive studies in Zambia, McGlashan has also shown a markedly coextensive distribution of diabetes and the practice of male circumcision. At first glance this seems a most curious relationship, suggesting possible surgical introduction of an unknown virus at an early age, and so forth. But after a bit more consideration, these data are quite compatible with the true scientific explanation: the higher prevalence of diabetes reflects a genetic fault in an inbred ethnic group.

"Etiology" and Contributory Causes. Many many illustrations could be given to emphasize the complex interplay between "etiology" and contributory causes. One will suffice here.

Hookworm infection, that is, ancylostomiasis, is a major cause of parasitic disease. The *infection* occurs over much of the Earth's land surface, and involves approximately 20 percent of the world's population—some 700 million persons! But hookworm *disease,* in a pragmatic sense, is quite a different matter—and the difference between infection and disease is not related merely to the number of infecting organisms, although this is an important factor. The principal disease manifestation of ancylostomiasis is anemia, a consequence of blood lost from the numerous minute penetrating wounds that the worms produce in the upper intestinal tract. Ordinarily, the bone marrow, with its enormous capacity to produce erythrocytes (given sufficient iron [and protein]), is quite capable of replacing, day by day, the blood lost as the consequence of even a heavy hookworm infection. Unfortunately, in many parts of the world, iron intake is barely sufficient to meet the demands of normal living. It is in these regions that available iron becomes the principle limiting factor in the body's efforts to make up for the continuing blood loss caused by the hookworms. And it is in these places where the diet is deficient in iron that we see most hookworm disease. In the Amazon River basin, for example, it is not uncommon for infants and small children to die as a result of hookworm disease. At the time of death their blood hemoglobin levels may be less than 10 percent of normal. The soil of the Amazon basin has been leached of most of its soluble iron, and the food, as well as the soil, is iron-deficient. Ethiopia provides a striking contrast. Although hookworm *infection* is common, hookworm *disease* is infrequent in most parts of the country, primarily because of the high iron intake of the Ethiopians. Interestingly, their source of iron is not the food, per se, but comes as a consequence of the primitive way in which the food is prepared. During the process of threshing teff, a wheatlike grain that is the staple of their diet, the minute grains become highly contaminated by the iron-rich soil. From these illustrations it is quite evident that an etiologic agent may require much assistance from contributory causes before it can produce overt disease.

The indirect is always as great and real as the direct.

WALT WHITMAN

HEREDITARY AND ENVIRONMENTAL FACTORS

Hereditary Factors

The genetic defect listed in Figure 2, **G6PD** (glucose-6-phosphate dehydrogenase) **deficiency,** handicaps the affected individual in a manner that is not directly apparent. This is because G6PD supports a metabolic mechanism that is ordinarily not essential because alternate mechanisms will meet normal requirements. (G6PD allows carbohydrate metabolism in the erythrocyte to follow the pentose phosphate pathway: glucose-6-phosphate—6-phosphogluconate.) This mechanism, however, becomes vital in protecting the erythrocyte from stress by certain oxidizing substances. The discovery of G6PD deficiency is a long and interesting story in itself. Suffice to say here that it was discovered as a by-product of research on antimalarial drugs. One of these drugs, primaquine (that acts as an oxidant), was found to produce hemolytic anemia in what first appeared to be a curious hit or miss fashion, affecting approximately 10 percent of Negroes, but virtually no Caucasians. It was no easy task to determine that a genetic defect was required to "prepare" the individual for injury by primaquine, and that this defect had a pronounced racial distribution (just as most of the hemaglobinopathies).

Skin color is also listed in Figure 2, and there are many ways that this genetically determined characteristic causally relates to disease. For example, since dark skin filters out a larger portion of carcinogenic actinic rays than does light skin, it protects from skin cancer—and Negroes have much less carcinoma of the skin than Caucasians. On the other hand, dark skin filters out a larger portion of those ultra violet rays that act to convert precursor substances in the skin to vitamin D—and children with dark skin are more likely to get rickets than those with light skin. But there are less obvious ways that skin color causally relates to disease. For example, the individual with a dark skin is at higher risk in those vast regions of Africa where sleeping sickness, African trypanosomiasis, is endemic. The apparent reason for this is that the vector of sleeping sickness, the tsetse fly, is more strongly attracted to dark than light skins, thus those with dark skins are more frequently bitten—and infected.

Under hereditary factors in Figure 2, **gastric secretion** is also listed. A normally high level of gastric acidity and enzyme activity certainly protects against infection from some kinds of microorganisms that use the intestinal tract as a portal of entry. But hypersecretion of HCl is a major factor in producing peptic ulcer. On the other hand, hyposecretion of HCl, more particularly achlorhydria (total lack of HCl secretion), a common condition among older age groups, contributes to the development of iron-deficiency anemia. Most dietary iron is in the ferric form. Before it can be used in the production of hemoglobin, it must be converted to the ferrous form; gastric HCl plays a major role in this conversion.

Environmental Factors

The influence of climatic factors is so well known that no example is required. A single illustration will suffice to point out the subtle influences of politico-economic and cultural factors. In the northern region of Nigeria, some of the inhabitants have, for many years, used a dark blue paste to color parts of their face and body, according to their custom. This material had been obtained from the French, and had an antimony base. With the changing politico-economic environment in the early nineteen sixties, the manufacturing source of the paste was also changed. Unfortunately, the new product contained significant amounts of lead. As a result, many individuals began to develop the typical manifestations of low-level lead intoxication, including the characteristic form of anemia (C. G. Berry, 1969, written commun.).

Geochemical Environment. There is a select, but unfortunately small group of diseases well known to be directly related causally to geochemical environment, for example: iodine deficiency and goiter, iron deficiency and anemia, fluoride intoxication and fluorosis. Rather than to discuss these, however, I wish to present specific illustrations of a much larger group of diseases, not so well known, many of which are *indirectly* influenced by geochemical environment.

Schistosomiasis, caused by *Schistosoma mansoni*[1], is a major disease in Brazil, but, fortunately, only in certain rather limited areas. Considering the primitive conditions that exist in the Amazon River basin, an area nearly two-thirds the size of Canada, one would think that this vast region would provide an ideal environment for schistosomiasis: temperature and humidity are nearly optimal; there is a great abundance of fresh water; and the habits of the people are such that there would be abundant opportunity for them to transmit the infection to the snail and, in turn, to have the snail transmit the infection back to them. But schistosomiasis does not exist in this area, except for two very small foci. The reasons for this virtual absence of the disease were not easy to discover. Finally, however, it became evident that those portions of the Amazon and its tributaries that fulfill the other ecologic requirements for schistosomiasis do not provide enough calcium to support the necessary snail intermediate host. Thus one of the links in the chain necessary to maintain schistosomiasis is broken. (As with most causal relationships, low calcium is not the *only* factor, but it is probably the most important. Another factor is the relatively low *p*H of the water, from humic acids. In some regions of the world the presence of copper and other heavy metals is thought to be

[1] Schistosomiasis represents a group of diseases caused by any one of three blood flukes: *Schistosoma mansoni, S. haematobium,* and *S. japonicum.* Each of these specific organisms produces its own form of disease, and each has its own geographic distribution, although they overlap.

These organisms have a complex life cycle, and each requires an intermediate host, a particular species of snail, in order to fulfill a vital stage of its development. The infected human being releases (excretes) eggs either in his feces or urine. If these eggs are discharged into fresh water, they hatch as free-swimming organisms (miricidia). Then, if they are able to reach the appropriate species of snail within a short time, they infect the snail (producing disease) and develop into a larval stage. At this stage they break out of the snail to become a free-swimming form once again (cercariae). If man is exposed to infested water for only a moment or two, the organisms are attracted to and can actively force entry through his skin and into cutaneous blood capillaries or venules. Once in the blood stream they are carried to their particular (organ) site of preference for further development. *S. mansoni* and *S. japonicum* prefer the veins which supply the colon, and the colon and rectum are primarily involved; *S. haematobium* concentrates in the blood vessels of the urinary bladder, and the bladder is the primary site of disease. As soon as the organisms develop to maturity they begin to lay eggs, and it is the eggs that cause most of the trouble, inciting a marked chronic inflammatory reaction. Major areas of primary direct damage (depending upon the type of organism) are the colon, the liver, and the urinary bladder. The kidneys, the lungs, and the heart become important sites of secondary, that is, indirect injury.

In most instances (most patients do not receive adequate treatment), the disease is a chronic one, lasting for many years. It is apt to become continually worse because continual re-exposure with re-infection adds continually to the number of infecting parasites.

Schistosomiasis is a very important cause of death, especially among infants. It is an even greater cause of morbidity because it depletes the energy of millions of people in the world, with very profound socio-economic consequences. Schistosomiasis may well be the world's most important disease!

Because of the complex life cycle of the schistosoma, that requires special conditions suited to the intermediate host, that is, a particular set of environmental circumstances allowing transmission of the disease from definitive host to intermediate host and back to definitive host, and so forth, schistosomiasis is one of the diseases that can be eliminated from an old area or introduced into a new area by changing the ecology of the area.

primarily responsible for the absence of the snail vector in fresh water that, otherwise, would seem to provide a suitable environment.)

Blackfoot disease has been recognized for many years in a small area along the western side of Taiwan, extending approximately 15 miles along the coast and 10 miles inland. This condition is one of dry gangrene. It has a prevalence as great as 0.5 percent and carries a mortality of 10 to 20 percent. Toes and feet are most commonly affected, but fingers and hands may be affected too. The condition first began to attract notice about 50 years ago and, looking backward, its development in the population seems to have followed an appropriate "incubation period." Fresh water is scarce in this part of China and, as the population increased, deep wells were dug in the area during 1900 to 1910. Some of these wells contained considerable arsenic—as much as 1.2 to 2 ppm. Cause and effect relationship between chronic arsenic poisoning and blackfoot disease is well established now, but it required a long time to make the connection. In addition to vascular effects and their sequelae, some 50 to 70 percent of blackfoot patients have arsenical dermatitis and, of this group, frank cancer of the skin occurs in approximately one among twenty.

Let us turn now to some diseases that may be causally connected with geochemical environment, but that require much more study before we can be certain.

The **Balkan nephropathy** is a good example. This endemic renal disease, which occurs in certain parts of Roumania, Yugoslavia, and Bulgaria, is characterized by progressive relentless destruction of the kidney—more specifically, atrophy and replacement fibrosis that affects predominantly the distal nephron. Characteristically, the disease makes a gradual asymptomatic onset, and by the time the condition is first recognized, the individual has advanced uremia. Curiously, arterial hypertension and edema are not prominent manifestations, as a rule. The usual course of the disease is six months to three years after it becomes apparent. According to one report, nearly 20 percent of the population of a particular region died of this condition during a period of 17 years. In one village, 30 percent of the inhabitants at a particular point in time were found to have proteinuria—the earliest manifestation of the disease. Epidemiologic studies strongly suggest that this is not an infectious disease nor the result of a genetic fault, despite its familial tendency. Cadmium, lead, nickel, and uranium have all, at one time or another, been considered as likely etiologic agents. The primary cause remains yet to be proved.

Amyotrophic lateral sclerosis (ALS) is another example of a disease that may reflect an unfavorable geochemical environment, perhaps a deficiency of manganese or magnesium. On the other hand, it may fall into the category of recently recognized "slow virus" disease. In any event, this progressive neurological disorder has a prevalence of more than 100 times the world average in a small area in Japan—the southern part of the Kii peninsula. Guam is another area of high prevalence. The people affected by ALS have a decreased intake of manganese and magnesium; they also have a decreased ability to store and use thiamin (vitamin B_1). They are benefited by medication that includes thiamin as well as magnesium and manganese.

There is some evidence that **Parkinson's disease,** a condition that strikes somewhat closer to home, may also be causally related to manganese, but in this instance, an excess of manganese. As with ALS, however, we need to learn much more before we can be sure of the etiology.

Many, many other diseases fall into this suspect category, diseases that appear to be strongly influenced by factors inherent in the geochemical environment—and that is one of the reasons why we are having this Symposium. Another reason focuses upon health, and it is important to realize that health is not merely the absence of disease. It is not a particular point on the plus side of the scale any more than disease is a particular point on the minus side of the scale. Health is a range that extends from buoyant to tolerable, and this Symposium is concerned with buoyant health as well as overt disease.

This Symposium is concerned also with that gray area between health and disease. When we think of sickness and its great cost, we must include those *subclinical* disease states which, though they do not ordinarily appear in the mortality or morbidity reports, nevertheless take a great toll, sapping both energy and motivation. As a consequence, the many individuals who are capable of working at half-normal efficiency or less cannot possibly do a day's work in a day's time. Furthermore, the individual's total work-life is reduced; a work-productive period of 15 or 20 years is common in many parts of the world. Even more important in the long run, their dampened motivation does not permit that necessary extra effort to become literate, or to improve manual skills, or to extend the quantity and quality of food produced, or to adopt effectively to a changing politico-economic environment. And this failure to improve contributes to those factors that were responsible for subclinical disease in the first place, thus the vicious cycle continues at the level of the individual, his family, and the community—and the state does not prosper—and the underdeveloped country does not develop.

SELECTED BIBLIOGRAPHY

Barton, P. A., 1955, The Natural History of Tsetse Flies: London, H. K. Lewis and Co., Ltd, p. 117.

Cheng Ch't, and Blackwell, R. Quentin, 1968, A controlled retrospective study of blackfoot disease: An endemic peripheral gangrene disease in Taiwan: Am. Jour. Epidermy, v. 88, p. 7–24.

Earle, K. M., 1968, Studies on Parkinson's disease including X-ray fluorescent spectroscopy of formalin fixed brain tissue: Jour. Neuropathology and Exp. Neurology, v. 27, p. 1–14.

Henschen, F., 1967, The History and Geography of Diseases (1962), transl. *by* Tate, J.: New York, Delacorte Press, 344 p.

Hopps, H. C., 1964, Principles of Pathology, 2d ed.: New York, Appleton-Century-Crofts, 403 p.

Kimura, K. and others, 1963, Epidemiological and geomedical studies on amyotrophic lateral sclerosis: Diseases Nervous System, v. 24, p. 1–5.

Kurkland, L. T., 1965, Amyotrophic lateral sclerosis: A reappraisal, *in* Gajdusek, D. C., Gibbs, C. J., Jr., and Alpers, M., *Editors,* Slow, latent, and temperate virus infections: Washington, U.S. Govt. Printing Office, Public Health Service Pub. 1378, p. 13–23.

May, J. M., 1958, The Ecology of Human Disease: New York, MD Publications Inc., 327 p.

Morita, S., Hattori, T., and Aoki, A., 1967, Medico-pedological studies on endemics: Chemical composition of soils in relation to amyotrophic lateral sclerosis in Kii Peninsula, Japan: Soil Sci. and Plant Nutr., v. 13, p. 45–52.

Stoll, N. R., 1947, This wormy world: Jour. Parasitology, v. 33, p. 1–18.

Weinstein, L., and Dalton, A. C., 1968, Host determinants of response to antimicrobial agents (with regard to glucose-6-phosphate dehydrogenase deficiency): New England J. Med., v. 279, p. 524–531.

Wolstenholme, G. E. W., and Knight, J., *Editors,* 1967, The Balkan Nephropathy: Boston, Little, Brown and Company, 123 p.

Yeh, S., 1965, Some geographic pathology aspects of common disease in Taiwan: Internat. Pathology, v. 6, p. 81–84.

——1964, The "Endemic Nephropathy" of South-Eastern Europe: Copenhagen, World Health Organization.

——1962, Tropical Health: A Report on a Study of Needs and Resources: Natl. Acad. Sci., Pub. 996, 540 p.

AUTHOR'S PRESENT ADDRESS: DEPARTMENT OF PATHOLOGY, UNIVERSITY OF MISSOURI MEDICAL CENTER, COLUMBIA, MISSOURI 65201

MANUSCRIPT RECEIVED BY THE SOCIETY JANUARY 8, 1970

Printed in U.S.A.

THE GEOLOGICAL SOCIETY OF AMERICA, INC.
MEMOIR 123, 1971

Chemical Compositions of Rock Types As Factors in Our Environment

HARRY A. TOURTELOT
U.S. Geological Survey, Denver, Colorado

ABSTRACT

The types of rocks that form geologic units in the Earth's crust supply most of the raw materials from which soils are formed and from which water derives its inorganic constituents. The compositions of what we eat and drink thus depend in part upon the compositions of the source rocks.

Igneous rocks are formed by crystallization and solidification of a rock melt. Metamorphic rocks are formed by recrystallization of both igneous and sedimentary rocks caused by heat and pressure within the Earth's crust. Sedimentary rocks are formed chiefly by the deposition in water of weathering and erosion products of pre-existing igneous, metamorphic, or other sedimentary rocks. The compositions of metamorphic rocks are generally similar to the compositions of the rocks that were metamorphosed, and only igneous and sedimentary rock compositions are considered here.

Igneous rocks range in SiO_2 content from about 40 to nearly 80 percent, and other constituents increase in amount as SiO_2 decreases. The changes in the other constituents are not large, however, except for the quantitatively unimportant least silicic rocks; these contain conspicuously more magnesium and less aluminum than the other kinds of igneous rocks.

Sedimentary rocks range in SiO_2 content from nearly zero for the carbonate rocks to almost 100 percent for quartzite and pure sandstone. Shale and clay contain intermediate amounts of SiO_2 and as much as 25 percent Al_2O_3, more than any of the igneous rocks. Carbonate rocks are composed mostly of calcium and magnesium carbonates.

The contents of individual trace elements vary widely with rock type. Chromium, titanium, nickel, and cobalt are conspicuously concentrated in low-silica igneous rocks that are quantitatively unimportant. Arsenic, iodine, molybdenum, and selenium are conspicuously concentrated in shale and clay.

In addition, most other elements occur in largest amounts in shale and clay compared to other sedimentary rocks, and in amounts nearly equal to those in igneous rocks.

Soils derived from different kinds of igneous rocks do not differ from each other as much as do soils derived from different kinds of sedimentary rocks. This is partly because igneous rocks generally are more resistant to weathering than sedimentary rocks that were deposited in water. Some of the important constituents of sedimentary rocks have been precipitated from solution, which makes them more susceptible to weathering and re-solution. Similarly, sedimentary rocks have a greater effect than igneous rocks on the composition of ground water.

Determination of areal variations in composition should be more detailed than normal petrologic investigations if the results are to be usable for environmental studies. Statistical principles should be used in planning the sampling, analysis, and interpretion of results.

CONTENTS

INTRODUCTION

The chemical compositions of the rocks that form geologic units in the upper part of the Earth's crust are fundamental geochemical factors in our environment. The rocks supply most of the raw materials from which soils are formed and from which water derives its inorganic constituents. Ultimately, the composition of what we eat and drink depends in part upon the compositions of the source rocks.

This paper summarizes data on the compositions of the types of rocks that make up the Earth's crust and the systematic relations in composition that exist among some of them. Such information is presented in an elementary way to facilitate the consideration by nongeologists of the potential chemical contribution of rocks to environmental characteristics. The regional surveys and detailed investigations necessary for understanding man's interaction with his environment can thus be placed in proper focus.

NATURE AND ORIGIN OF TYPES OF ROCKS

The rocks that make up the Earth's crust are classified into three major groups—igneous, sedimentary, and metamorphic (Fig. 1). The three types are interrelated in a complex system that has been operating throughout discernible geologic time but which can be regarded conveniently as starting with a rock melt. From this melt, the rocks called "igneous" are crystallized, obeying the laws of chemistry applicable to the composition of the melt and to the conditions of temperature and pressure controlling the process. Sometimes crystallization or glassy solidification takes place at the Earth's surface, as when a volcano pours out lava, which is simply a rock melt. Once formed and exposed on the Earth's surface, igneous rocks are subject to weathering and produce soluble and residual chemical and detrital products. These products tend to move toward the oceans and, with the addition of some biologic materials, form sedimentary rocks. Although most sedimentary rocks are formed in oceans, many are formed in nonmarine situations along the drainage ways by which the sediments are removed from their source.

Both igneous and sedimentary rocks can be brought, by geologic processes, from the Earth's surface into lower regions of the crust where conditions of heat and pressure cause either chemical or physical rearrangement of the constituents, or both, so that a different kind of rock is formed. This process is called metamorphism and the resulting rocks are called "metamorphic."

Obviously, all three types of rocks are exposed at the Earth's surface and all three can be the raw material for the formation of additional sedimentary rocks. It should also be obvious that all three types can be subjected to metamorphism to produce additional kinds of metamorphic rocks. Present understanding of the processes operating in the Earth's crust also allows for all three types of rocks to be brought into conditions of temperature and pressure sufficient for a new rock melt to be formed and for the cycle to start over again.

Proportions on Earth's Surface

If recognizable metamorphosed sedimentary rocks are classified as sedimentary rocks and metamorphosed igneous rocks are classified as igneous,

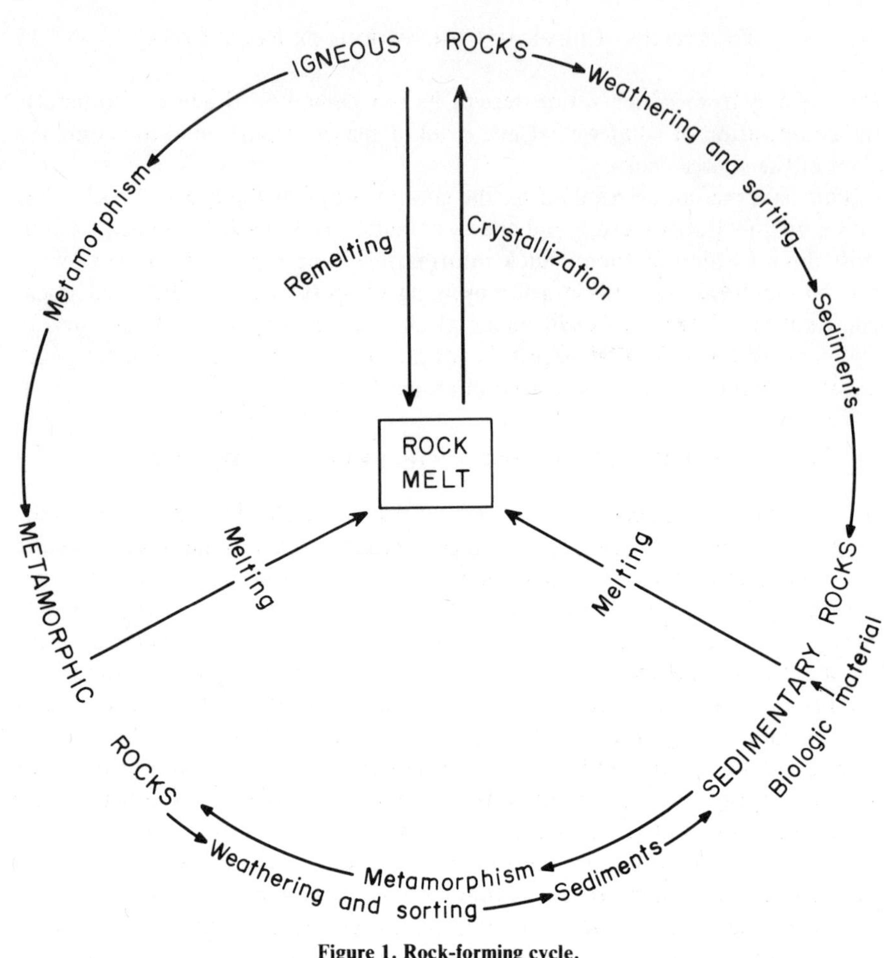

Figure 1. Rock-forming cycle.

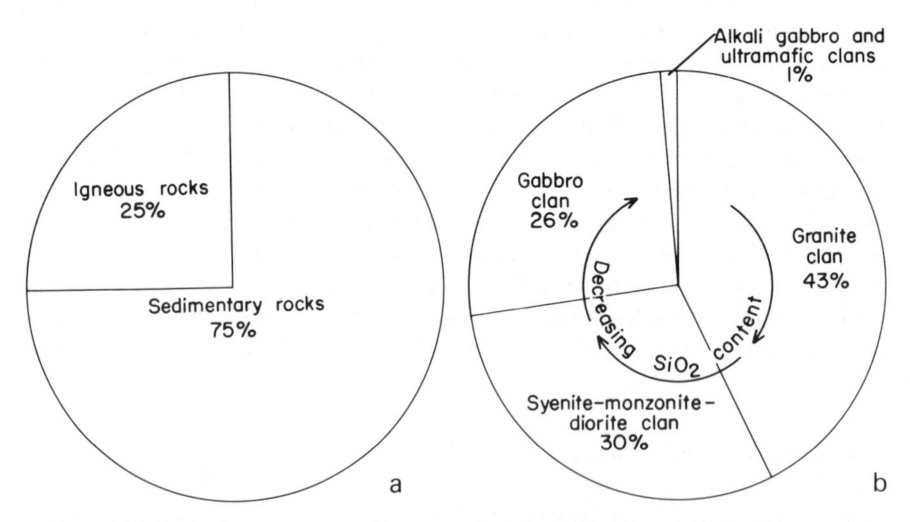

Figure 2. Relation of outcrop areas of igneous and sedimentary rocks and kinds of igneous rocks. (a) Area of outcrop. (b) Areas of outcrop of kinds of igneous rocks.

then about 75 percent of the Earth's surface is underlain by sedimentary rocks and 25 percent by igneous rocks (Pettijohn, 1957, p. 8) (Fig. 2). Inasmuch as the chemical composition of metamorphic rocks generally is similar to that of source rocks, this combining is justifiable to arrive at the above figures. In addition, however, for the purposes of this paper it allows us to ignore metamorphic rocks as a rock type—and from here on, only igneous and sedimentary rocks will be discussed.

CLASSIFICATION OF IGNEOUS ROCKS

The classification of igneous rocks is a complicated subject that depends upon a combination of mineralogical and chemical data, and the mineralogical aspects of classification cannot be treated here. From the several points of view that can be applied to the problem, the approach used by Williams and others (1954) has been selected. Igneous rocks, in order of decreasing silica content, are placed in five clans: the granite, syenite-monzonite-diorite (hereafter referred to as the syenite clan), gabbro, alkali-gabbro, and ultramafic clans. The relative areas of North America underlain by these different rock types as estimated by Daly (1914, p. 44) also are shown on Figure 2. Rocks of the granite clan compose 43 percent of the area underlain by igneous rocks; rocks of the syenite and gabbro clans make up the remainder and are about equally divided. The alkali-gabbro and ultramafic clans are quantitatively unimportant.

Average Major-Element Compositions of Igneous Rocks

The compositions of these rocks are shown in Table 1 and on Figure 3. The data are derived from 134 analyses of igneous rocks presented by Clarke (1924, p. 439–470). Clarke's analyses are used, first, because it is suspected that they were carefully selected by Clarke to represent typical examples of compositional groups utilized in a particular classification scheme (*see* Clarke, 1924, p. 423–437) and, second, because the data permit the ready determination of means and standard deviations by computer techniques. Measures of dispersion around the mean are lacking in better known averages of igneous rock compositions such as those of Daly (1914; 1933) and Nockolds (1954). Before the means and standard deviations were computed, the minor constituents were eliminated from Clarke's analyses and the remaining major constituents were recalculated to 100 percent.

The mean for each constituent, with a convention showing the standard deviation, is plotted for each clan against the mean silica content and the standard deviation of silica for that clan. The great overlap of the standard deviations for silica indicates that the classification of igneous rocks depends only partly upon the chemical composition of the rock. If the rocks are regarded as the result of a chemical system, however, the classification of igneous rocks reflects a transition from granitic rocks containing about 75 percent silica at one end of the scale to ultramafic rocks containing less than 50 percent silica and substantial amounts of calcium, magnesium, and iron at

TABLE 1. AVERAGE COMPOSITION OF IGNEOUS AND SEDIMENTARY ROCKS

	SiO_2		Al_2O_3		Fe_2O_3		FeO	
	$\bar{x}$	s	$\bar{x}$	s	$\bar{x}$	s	$\bar{x}$	s
Igneous Rocks								
Granite clan	75.43	1.85	13.65	1.16	0.76	0.41	0.87	0.60
Syenite-monzonite-diorite clan								
Syenite	60.52	5.86	16.39	2.66	2.62	1.46	2.58	1.97
Monzonite	62.10	4.94	16.63	1.37	2.94	0.92	2.87	1.13
Diorite	59.50	5.95	16.58	1.53	2.65	1.19	4.54	2.81
Alkali-gabbro clan	53.64	6.27	17.99	5.30	3.73	1.68	2.77	1.91
Gabbro clan	52.19	3.86	16.93	3.72	2.68	1.37	7.59	3.68
Ultramafic clan	47.32	5.12	3.61	2.72	3.47	2.59	8.02	3.57
Average of all analyses	58.41	9.89	15.03	5.39	2.71	1.75	4.09	3.48
Sedimentary Rocks								
Sandstone	92.12	6.46	2.12	1.78	1.08	1.20	—	—
Sandy shale and clay	68.44	8.11	14.29	3.92	3.66	1.74	1.69	1.52
Shale and clay	56.88	6.82	25.22	7.07	3.73	3.14	6.75	—
Calcareous sandstone	46.40	18.62	2.32	1.82	1.54	1.20	1.38	0.93
Calcareous shale and clay	27.40	10.69	8.08	3.81	2.84	1.40	1.30	1.03
Carbonate rocks	4.85	4.38	1.29	1.13	1.20	1.05	0.67	0.66

Analyses in percent. Igneous rock averages calculated from 134 analyses selected by Clarke (1924, p. 439–470). Sedimentary rock averages calculated from 2425 analyses compiled by Hill

the other end. This narrow range of silica content, in comparison with that for sedimentary rocks, results from the fact that most of the minerals in igneous rocks are silicate minerals, and there must always be some minimum amount of silica in a rock.

The changes in amounts of the various constituents as the silica content decreases are not large and are reasonably systematic between granite and gabbro. The relatively large amounts of sodium and potassium in syenites and alkali gabbros are exceptions to these trends. The ultramafic rocks differ sharply in composition from the other rock clans.

In general, it can be anticipated that most virgin soils formed from any of the igneous rocks except ultramafic ones will be normally productive. Poorly drained soils might be undesirably alkaline if formed on syenites or on the rare alkali gabbros. The composition of a virgin soil on ultramafic rock, however, probably would reflect the low sodium and potassium contents of the parent rock and would contain an undesirable excess of magnesium and iron. Although ultramafic rocks are rare, even a small area of exposure could be a significant contaminating factor for soils and waters, depending upon the geologic situation that controls the distribution of the products of weathering and decay.

TABLE 1. (CONTINUED)

MgO		CaO		Na_2O		K_2O		CO_2		Number of analyses
$\bar{x}$	s	$\bar{x}$	s	$\bar{x}$	s	$\bar{x}$	s	$\bar{x}$	s	
Igneous Rocks										
0.28	0.22	0.92	0.63	3.21	0.95	4.88	1.27	—	—	21
3.17	3.81	4.25	2.80	5.50	2.72	4.98	2.10	—	—	16
2.50	1.75	4.80	2.08	3.63	0.64	4.52	0.41	—	—	8
4.13	2.72	6.35	2.66	3.65	0.97	2.60	1.12	—	—	22
3.63	4.34	5.28	4.08	7.22	3.90	5.74	2.57	—	—	31
7.46	3.45	9.13	2.91	2.70	1.06	1.33	1.02	—	—	21
30.77	9.66	6.10	4.65	0.39	0.51	0.33	0.63	—	—	15
6.70	9.83	5.32	3.91	4.11	3.05	3.63	2.53	—	—	134
Sedimentary Rocks										
0.32	0.40	0.68	0.86	0.62	0.54	0.48	0.56	2.57	—	101
1.48	1.48	2.04	2.20	1.61	—	2.49	1.53	4.31	3.43	839
1.33	1.29	0.98	1.17	1.27	1.31	2.03	1.73	1.80	1.69	294
3.49	3.77	23.05	10.53	0.66	0.33	1.41	0.92	19.74	8.36	85
3.38	4.00	30.10	10.57	0.75	0.58	1.48	0.63	24.66	7.17	97
4.52	6.87	46.53	8.54	0.17	0.20	0.28	0.26	40.51	5.29	1009

and others (1967). The selected constituents have been calculated to 100 percent. $\bar{x}$ = arithmetic mean; s = standard deviation.

Water derived from igneous rock terranes generally contains relatively few dissolved chemical constituents because most minerals in igneous rocks are only slightly soluble in water and because igneous rocks have little porosity and hence contain little water to attack the minerals.

NATURE AND ORIGIN OF KINDS OF SEDIMENTARY ROCKS

The sedimentary rocks, in contrast to the igneous rocks, are the result of a mixture of mechanical and chemical processes. The chief components are sand, clay, and carbonate (Fig. 4). Sand forms sandstone and clay forms shale and claystone when these constituents are consolidated into sedimentary rocks.

"Sand" refers to the particles produced by the mechanical weathering of some pre-existing rock. In the broadest possible sense, the term "sand" can refer to particles of any size, although special names are used for very large and very small particles. Sand can be made up of any mineral or group of minerals present in the source rock, but most sands consist chiefly of the mineral quartz, which is nearly pure silica. Silica is relatively insoluble under the conditions at the surface of the Earth, so it endures long after the other

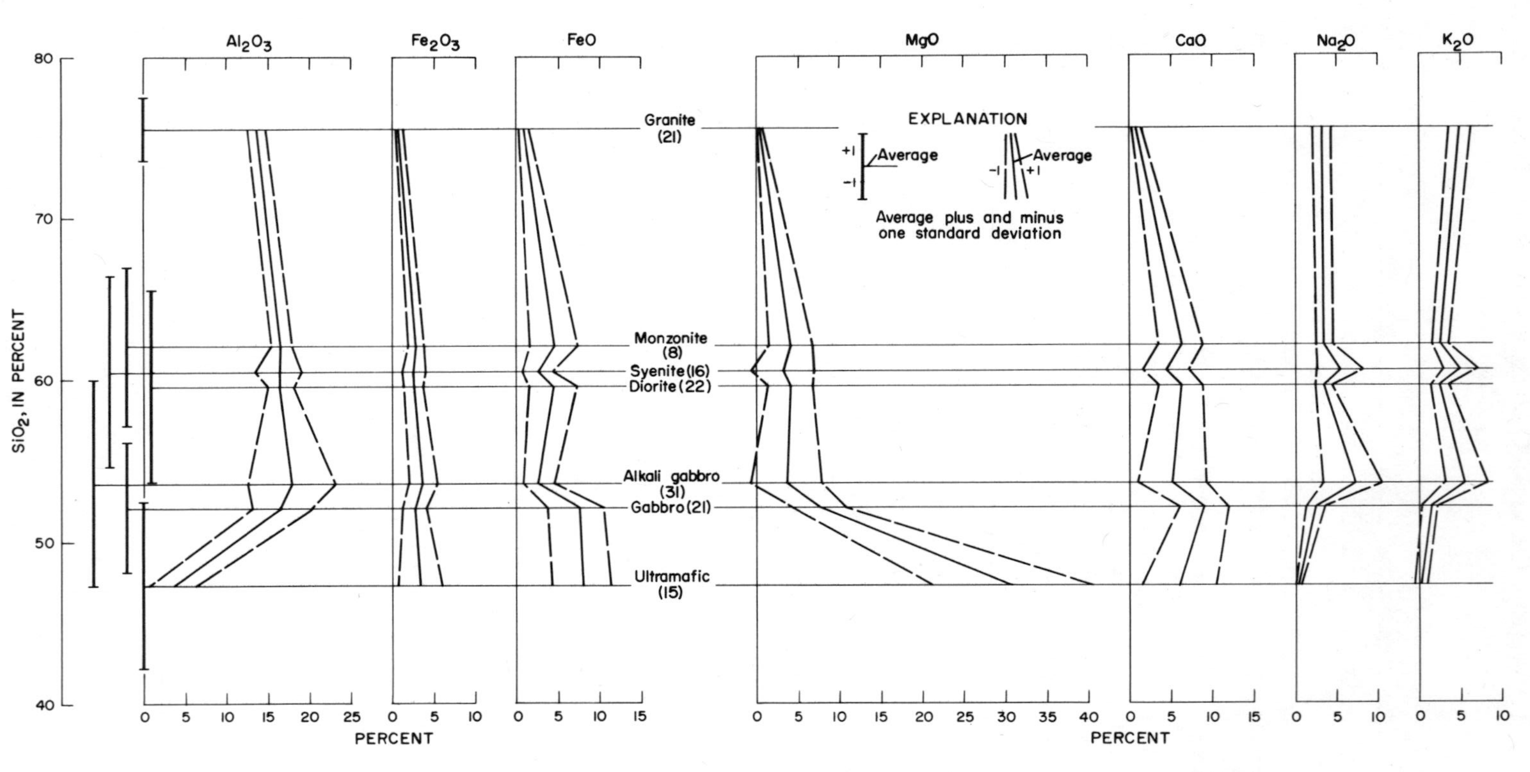

Figure 3. Average contents of major constituents in types of igneous rocks. Based on 134 analyses (Clarke, 1924, p. 439–470).

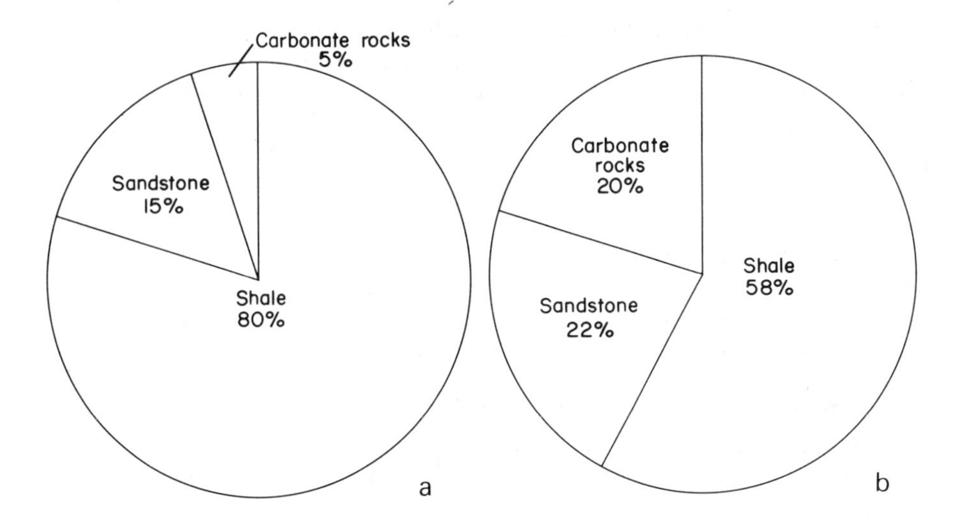

Figure 4. Volumetric and areal proportions of kinds of sedimentary rocks. (a) Calculated volumes in Earth's crust. (b) Estimated proportions from thickness measurements.

less resistant minerals have disappeared under ordinary conditions of weathering. In addition, quartz is the most abundant mineral in the rocks of the granite clan and thus would be expected to be prominent in the sediments and sedimentary rocks derived from this source. Sand is distributed on the Earth's surface by flowing water or wind.

Clay—using the word as a general term—is made up of layer-structured aluminum silicate minerals in which alkalies and alkaline earths are the principal cations. Most clay results from the weathering of aluminum silicate minerals in the source rock, but some results from the combining of ions released by the weathering of glassy, poorly crystalline igneous rocks of the volcanic type. Once formed, clay is distributed mechanically by the same processes as sand.

A distinctive characteristic of clay is that it occurs in very small particles ranging from several microns in maximum dimension down to much less than 1 micron. The layer structure of clay minerals combined with the colloidal size of many if not most of the typical particles makes clays particularly susceptible to exchange and adsorption reactions. The cations are held loosely in and on the fundamental layer structure, and they can be exchanged easily as the clay particle moves from one chemical environment, such as fresh water, into another, such as the ocean. In addition, defective or broken layer structures can permit binding of heavier metallic elements.

"Carbonate" refers to limestone (calcium carbonate) and dolomite (calcium-magnesium carbonate). Limestone is formed most commonly by a mixture of mechanical and chemical processes. Mechanical processes include the accumulation of the shells of microscopic animals, broken fragments of larger shells, and eroded coral, resulting in particles ranging from sand to clay in size. Chemical processes include direct precipitation of calcium carbonate under appropriate conditions either as beds or as cement that binds together the particles of mechanical origin. Dolomite may originate by direct precipitation under rarely occurring conditions but most commonly is the result of replacement processes by which magnesium is added to a previously existing calcium carbonate.

Additional components of sedimentary rocks are the chemical precipitates of chlorides (such as rock salt) and sulfates (such as gypsum and anhydrite). These components are important natural resources and they can have marked effects on the environment at localities where they are exposed. Pure chloride and sulfate deposits bulk so small, however, among the sedimentary rocks as a whole that they are not considered further here.

Proportions of Kinds of Sedimentary Rocks

The estimates of the abundance of sedimentary rock types shown in Figure 4 are gross estimates concerned with pure end members only. The calculated proportions given by Clarke and Washington (1924, p. 32) reflect chiefly relative crustal volumes, and sedimentary rocks as a class amount to only 5 percent of the Earth's crust. The estimate (Pettijohn, 1957, p. 10) based on thicknesses of the rock types exposed on the Earth's surface should approximate

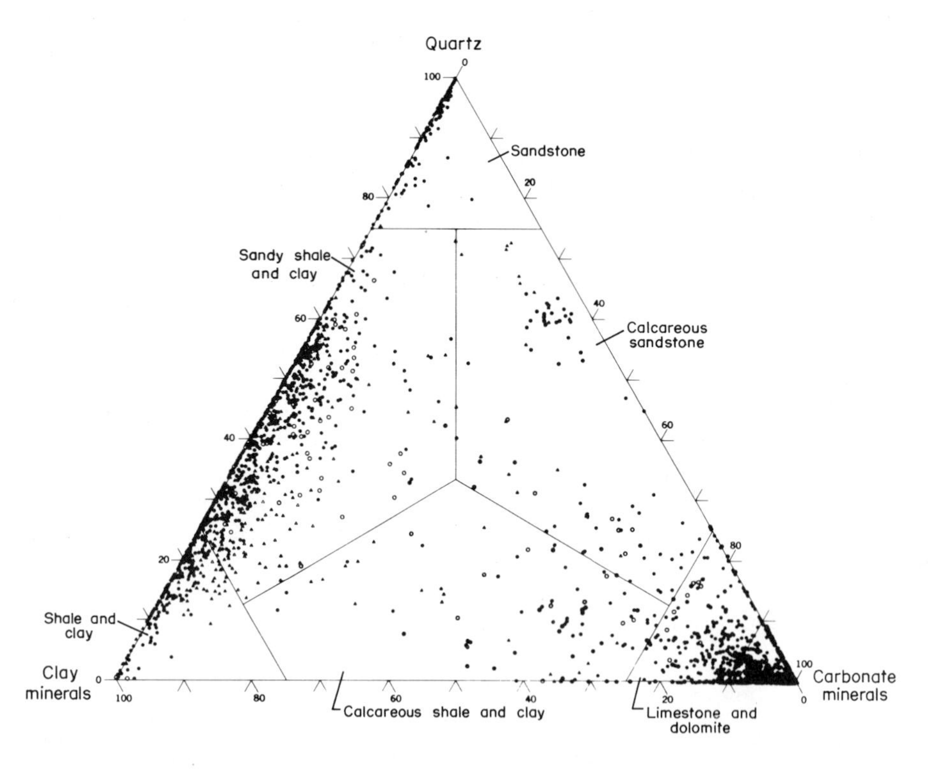

Figure 5. Proportions of quartz, clay minerals, and carbonate minerals calculated from about 2400 analyses compiled by Hill and others (1967).

the relative outcrop areas of the rock types. Sandstone and limestone, being harder than clay or shale, impede erosion and thus tend to hold up large areas at the surface. Both of these rock types have additional importance in environmental geochemistry because they provide the aquifers from which most ground water is obtained. Because of the fine particle size, clay lacks the permeability to be an effective aquifer.

Figure 5 illustrates how quartz, clay minerals, and carbonate minerals combine to form sedimentary rocks. The data consist of about 2400 analyses compiled by Hill and others (1967) for the Great Plains states. Most of the samples represented are mixtures of sand and clay or else relatively pure carbonate rocks. Samples representing subequal mixtures of all three components are few. The published analyses that were compiled were made chiefly for economic purposes. Far too many limestone analyses are included relative to those of other rock types for the distribution to reflect the abundance of these rocks in nature. Limestones and sandy clays are more often analyzed than other rocks because chemical data are necessary to judge their economic worth.

The area of the triangle in Figure 5 is divided into six fields: three represent end members containing more than 75 percent of the indicated constituent and three represent mixtures between pairs of the constituents.

Average Major-Element Compositions of Sedimentary Rocks

The compositions of these sedimentary rocks as indicated by the analyses compiled by Hill and others (1967) are shown in Table 1 and Figure 6. As mentioned, the rock names for which compositions are plotted are defined primarily on the basis of composition. Consequently, the overlap of the standard deviations for silica for each rock name is smaller than that in the igneous rocks. The overlap between shale and sandy shale reflects the composition of different clay minerals in the rocks. The range in silica content is much greater among sedimentary rocks than among igneous rocks: a sandstone of nearly pure quartz can be almost 100 percent silica, and, at the other end of the scale, a pure limestone or dolomite is made up of carbonate minerals and may contain essentially no silica.

The curve for aluminum oxide illustrates how sand and clay form a mixing system, with the clayey rocks containing the most aluminum. The curve for calcium oxide illustrates the mixture of carbonate minerals with both sand and clay, the carbonate rocks containing the most calcium and calcareous sandstone containing the least. The clays contain the most ferrous oxide; ferrous sulfide (pyrite) and ferrous carbonate (siderite) are common minerals. The minerals form shortly after deposition of the clays if there is sufficient organic matter mixed with the clay for the oxygen trapped in the clay to be exhausted by the decay of the organic matter. The clays and sandy clays contain about equal amounts of potassium and sodium. In the clays, these alkalies are in clay minerals, but in the sandy clays, some of the alkalies must be in feldspars that are part of the sand fraction to account for the high alkali

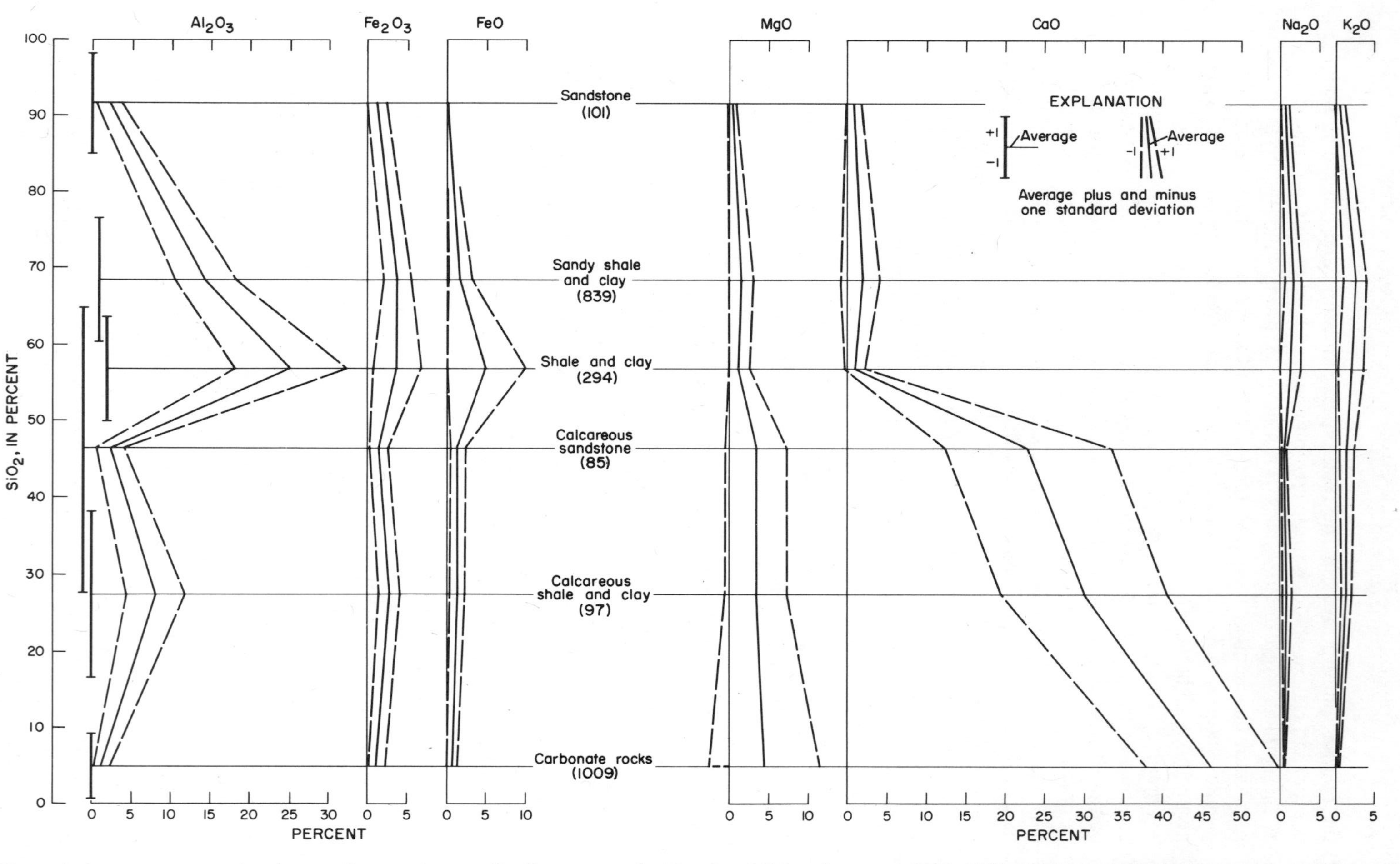

Figure 6. Average contents of major constituents of types of sedimentary rocks. Based on 2425 analyses compiled by Hill and others (1967). Symbols same as on Figure 3.

content. The relatively low magnesium content of the carbonate rocks may reflect the general scarcity of dolomitic rocks in the area represented by the analyses as well as the small number of dolomite analyses included in the data.

The shifts in composition for most of the constituents, as the silica content decreases, are much larger for sedimentary rocks than for the igneous rocks. This seems to be because the constituents of sedimentary rocks, for the most part, are brought together by mechanical processes that are more highly varied than the physical constants of temperature and pressure under which igneous rocks crystallize or solidify. These large differences in composition between kinds of sedimentary rocks account for the fact that soils and waters derived from sedimentary rocks also vary in composition. A soil derived from syenite might not be readily distinguished from a soil derived from granite, but a soil derived from sandstone is quite different from one derived from a thick stratum of clay.

Sedimentary rocks contain more minerals that are soluble under near surface conditions or readily alterable than are contained in igneous rocks because many of the minerals in sedimentary rocks were precipitated from aqueous solution. Consequently, sedimentary rocks have a greater effect on the composition of water than igneous rocks generally have. In addition, sedimentary rocks form the chief aquifers for ground water. Most water flowing in streams has previously passed through some kind of ground-water reservoir. Such a reservoir might be a regionally deep one that local geologic structure brings to the surface so that artesian conditions exist and springs are formed. Or the ground-water table might be intersected by the ground surface so that water can discharge. Or the reservoir might simply be the soil and other surficial deposits on hillsides in which rainfall is adsorbed and gradually released at lower topographic positions. The conditions under which the water both enters and leaves these reservoirs offer the opportunity for unstable minerals to be altered and the composition of water to be affected.

Ferrous sulfide readily oxidizes to ferric oxide and sulfuric acid. If calcium-bearing materials are available, the sulfuric acid reacts with them to form calcium sulfate (gypsum) that can contaminate water for many miles downstream from the source. The ferric oxide usually is deposited fairly close to the site of its formation, but it too can be a strong local contaminant of water. If the clays in which the ferrous sulfide occurs have sodium as their principal cation, then sodium sulfate or alkali is formed with deleterious effects on both soils and waters. Any of these substances can be concentrated in water or soils by simple evaporation and hence may be more serious problems in arid than in humid regions.

AVERAGE TRACE-ELEMENT COMPOSITIONS OF IGNEOUS AND SEDIMENTARY ROCKS

Metallic elements present in source rocks in small amounts—the so-called minor elements or trace elements—have been shown to have important effects on human and animal health, resulting from their excesses or deficiencies in soils, waters, and plants. Average amounts of these elements in specific rock types based on uniform bodies of data are not available. Estimates based on all data must be used and no limits to the uncertainty can be assigned. The data for some trace elements taken from Turekian and Wedepohl (1961) are shown in Figure 7. These elements generally occur in rocks in amounts ranging from about 100 to 1000 ppm and include phosphorus, sulfur, and manganese, all of which are important biochemical elements. The other elements in this group also have recognized biochemical roles. The variation in amount from rock type to rock type is different for each element but commonly amounts to ten-fold or more and for some elements, such as barium, may be even larger. Systematic relations between trace-element content and rock type are similar to those of some of the major constituents. Gabbro, containing about 50 percent silica, has the largest amounts of phosphorus, strontium, titanium, and vanadium and is not markedly deficient in any elements. Shale (about 60 percent silica) and limestone (5 percent silica) contain more sulfur than any other rock type. Among the sedimentary rocks, shale contains more of all elements except strontium, which is present in largest amount in limestone, and zirconium, which is present in largest amount in sandstone (about 80 percent silica).

The data for another group of elements with recognized biochemical roles ranging from beneficial to deleterious are shown in Figure 8. These nine elements occur in rocks in amounts generally smaller than 100 ppm. Among the sedimentary rocks, shale contains the largest amounts of all nine elements. Shale is the only rock type that contains significant amounts of selenium. Cobalt generally is not present in detectable amounts in limestone. The average contents for both nickel and cobalt are largest in the rare ultramafic rocks (about 45 percent silica). Gabbro (about 50 percent silica) contains as much or more of all these elements as the other igneous rocks.

All the minor elements find their way into soils and waters by the same paths mentioned for some of the major elements (*see* p. 18 and 25). Because many of the minor elements are located in the source rocks in relatively unstable minerals, the dissemination of the minor elements may be more extensive than for the major elements. Some of the more biochemically critical elements, such as copper, molybdenum, selenium, and zinc, are readily adsorbed by various organic compounds and may be concentrated in organic-rich soils far above the amounts found in the source rocks from which those soils were formed. Some plants also have the ability to concentrate elements in their tissues in larger amounts than are found in the soil or the parent bedrock, and the decay of these tissues may lead to further enrichment of the soils.

VARIATIONS IN COMPOSITIONS

So far, only average amounts of all the elements have been discussed. Amounts of some elements may be much larger in certain varieties of the broad rock types considered here. One of the most notable examples is the

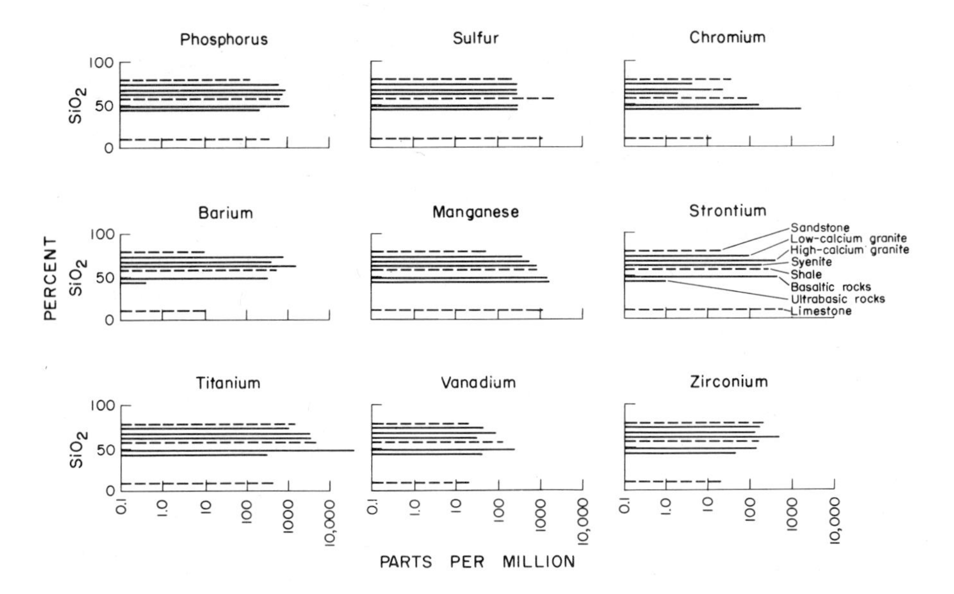

Figure 7. Average contents of minor elements occurring in the general range 100 to 1000 ppm in igneous (solid lines) and sedimentary (dashed lines) rocks. Rock types plotted against silica content. All data *from* Turekian and Wedepohl (1961). The names of some of the igneous rock types are different from those used in text and Figure 3 but represent the same sequence of decreasing silica content.

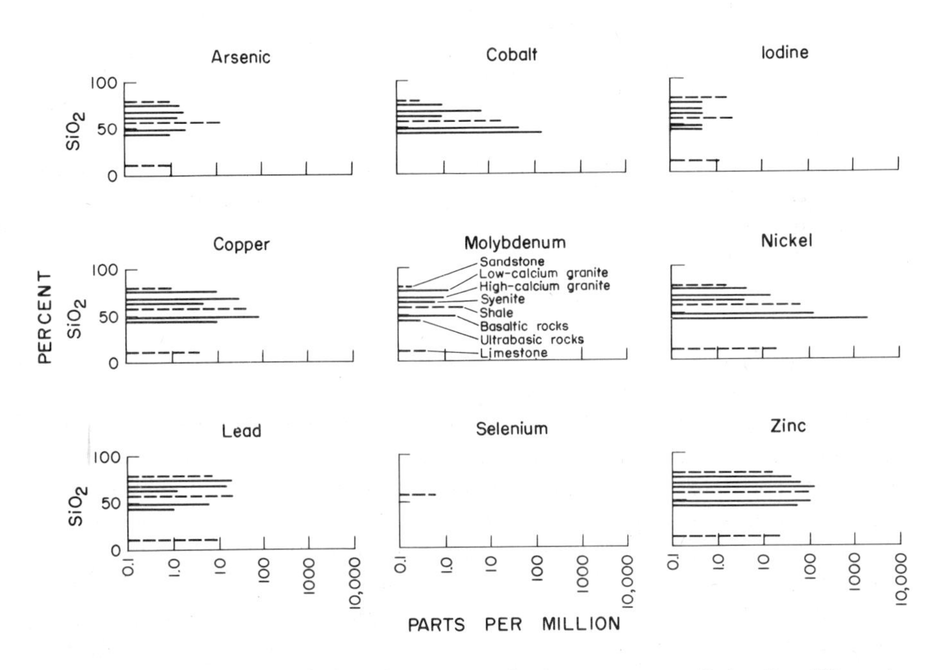

Figure 8. Average contents of minor elements occurring in amounts generally less than 100 ppm in igneous (solid line) and sedimentary (dashed line) rocks. Rock types plotted against silica content. All data *from* Turekian and Wedepohl (1961). The names of some of the igneous rock types are different from those used in text and Figure 3 but represent the same sequence of decreasing silica content.

concentration of many of the minor elements in organic-rich varieties of shale. Krauskopf (1955) has reviewed the range of amounts of metallic elements that can be found in organic-rich shales. Arsenic, copper, molybdenum, and zinc can occur in such rocks in amounts larger than those shown in Figures 7 and 8 by factors of five to several thousand. Inasmuch as organic-rich shales usually have a large areal extent, such shales are important environmental factors. Many of the biologic problems presented by excesses of selenium, molybdenum, and other metals in soils are related to such organic-rich shales.

In addition, however, the contents of both major chemical constituents and minor elements can vary within the areal extent of a given rock type. Within an extensive terrain of basalt (a member of the gabbro clan), such as the Columbia Plateau in parts of Washington, Oregon, and Idaho, the magnesium content, for example, may differ markedly from place to place. Some of the differences may be related to the different times at which different basalts were extruded. Parts of an individual flow may vary in chemical composition as a result of differential cooling or of differentiation of parent magma material in the lava chamber. The minor-element content of a sandstone unit also can change areally. In general, more clay is deposited in a sand as distance from the source increases, and the additional clay will contribute additional minor elements.

This paper has presented a very much simplified view of rock compositions and their effects on soils and waters. The composition of a soil cannot be predicted precisely from the fact that it lies within an area shown on geologic maps to be a certain rock type, because: (1) the rock type may depart considerably from the average in its composition, and (2) soil-forming processes can be highly varied even within the same climatic regimen and uniform source rock. The compositions of waters can depart from expectable averages for similar reasons.

FURTHER INVESTIGATIONS

Both regional and detailed geochemical surveys are needed in order to determine areal variation in rock compositions with sufficient precision for the data to be meaningfully related to health/disease maps if we are to correlate these conditions with geochemical environment. Such surveys would differ from most made previously in that they would require far more data points than ordinarily are used in petrologic studies. This would require the use of statistical tools in the interpretation of the data, which has the further requirement that the sampling design should be appropriate for the interpretive techniques contemplated. Only by this approach will the geochemical data and the interpretations of these data be comparable with existing demographic and epidemiologic information. In this way, chemical studies can contribute additional dimensions to the study of health and disease.

REFERENCES CITED

Clarke, F. W., 1924, The data of geochemistry [5th ed.]: U.S. Geol. Survey Bull. 770, 841 p.

Clarke, F. W., and Washington, H. S., 1924, The composition of the Earth's crust: U.S. Geol. Survey Prof. Paper 127, 117 p.

Daly, R. A., 1914, Igneous rocks and their origin: New York, McGraw-Hill Book Co., Inc., 563 p.

——1933, Igneous rocks and the depths of the earth: New York, McGraw-Hill Book Co., Inc., 598 p.

Hill, T. P., Werner, M. A., and Horton, M. J., 1967, Chemical composition of sedimentary rocks in Colorado, Kansas, Montana, Nebraska, North Dakota, South Dakota, and Wyoming: U.S. Geol. Survey Prof. Paper 561, 241 p.

Krauskopf, K. B., 1955, Sedimentary deposits of rare metals, p. 411–463, *in* Bateman, A. M., Econ. Geol., 50th anniversary volume, 1905–1955, pt. 1: Urbana, Ill., Econ. Geol. Pub. Co.

Nockolds, S. R., 1954, Average chemical compositions of some igneous rocks: Geol. Soc. America Bull., v. 65, no. 10, p. 1007–1032.

Pettijohn, F. J., 1957, Sedimentary rocks [2d ed.]: New York, Harper & Brothers, 718 p.

Turekian, K. K., and Wedepohl, K. H., 1961, Distribution of the elements in some major units of the Earth's crust: Geol. Soc. America Bull., v. 72, no. 2, p. 175–192.

Williams, Howel, Turner, F. J., and Gilbert, C. M., 1954, Petrography: An introduction to the study of rocks in thin sections: San Francisco, W. H. Freeman & Co., 406 p.

PUBLICATION AUTHORIZED BY THE DIRECTOR, U.S. GEOLOGICAL SURVEY

MANUSCRIPT RECEIVED BY THE SOCIETY JULY 10, 1969

Printed in U.S.A.

THE GEOLOGICAL SOCIETY OF AMERICA, INC.
MEMOIR 123, 1971

Regional Geochemical Reconnaissance in Medical Geography

JOHN S. WEBB
Imperial College, London, England

ABSTRACT

Geochemical reconnaissance by stream-sediment sampling provides a ready means for mapping the distribution of the elements on a regional scale. The method, which is rapid and inexpensive, is based on the premise that stream sediment approximates to a composite sample of products of weathering and erosion of the rocks and soils upstream from the sample site. Although the technique is simple, interpretation of the data contains a strong element of research.

Geochemical reconnaissance of this type is now standard practice in mineral exploration. Recently the results of trial surveys carried out in the United Kingdom at a sampling density of 1 sample per sq mi, followed by analysis for some twenty elements, have demonstrated a useful range of application in agriculture as a means of delineating suspect areas of trace-element disorders in crops and animals. The significance of stream-sediment data in medical geography is a matter for interdisciplinary research. Nevertheless, there are grounds for believing that a geochemical atlas could provide a unique source of ancillary information in epidemiological surveys, particularly in rural areas and in the less well-developed parts of the world.

CONTENTS

Figure

Table

INTRODUCTION

While the importance of trace elements in nutrition is generally accepted, there remain considerable areas for debate concerning the importance of geology in medical geography. This is true even of diseases known to have trace-element affiliations. Controversy can be particularly acute when considering the relation between geology and disease in highly developed communities and particularly those diseases concerning which medical opinion is itself divided or in doubt as to the causal factors. It is not the purpose of this paper to discuss the controversial epidemiological issues, but rather to indicate one way in which the geologist-geochemist may usefully contribute to the study of these important and truly interdisciplinary problems.

Probably the greatest twin obstacle facing an objective assessment of the scope for geology in epidemiology is the lack of adequate medical-geographical statistics, on the one hand, and of an atlas showing the regional distribution of the elements, on the other. I am concerned here only with the latter. It is doubtful whether anyone would dispute the value of a geochemical atlas. The problem lies in the method by which it should be compiled. The geological map is, of course, an essential prerequisite but by itself gives no more than a valuable guide to the probable ranges of concentration of the elements, which can vary significantly even within a single rock formation. Detailed surveys of the elements in rocks, soil, vegetation, water, and dietary intake are too costly and time-consuming to be contemplated on a regional scale. Regional conclusions extrapolated from small bodies of type-locality data are generally unconvincing, if the geochemical variations that can exist within geological formations and the profound influence that relatively minor changes in the surface environment can exert on the geochemical behavior and distribution of the elements are borne in mind. It is contended that the answer lies in rapid low-cost geochemical reconnaissance aimed at coverage for upward of 20 elements at rates of the order of 20,000 sq mi per team-year, with provisional maps produced annually for immediate user consumption.

Regional reconnaissance to this specification is, of course, no substitute for detailed surveys. Rather, it is a natural prelude providing, quickly and economically, an over-all picture of trace-element distribution. This advance information, used in conjunction with geological, pedological, and other appropriate maps, would then serve as a valuable aid in selecting worthwhile problem areas and optimum programs for subsequent investigations in detail.

STREAM-SEDIMENT RECONNAISSANCE

Research over the past fifteen years has shown that stream-sediment sampling is a practical proposition in geochemical reconnaissance. The method is based on the premise that the products of weathering are funnelled down the surface drainage system. For most purposes, the sediment on the stream bed is preferred as the drainage sampling medium because of fluctuations in the metal content of waters related to climatic and other secondary factors and the greater difficulties of transporting and analyzing water samples. In any event, as a result of transfer between water and sediment, the composition of the latter can often give a guide to variations in both relatively soluble and insoluble constituents. By virtue of its origin, therefore, stream sediment approximates more or less closely to a composite sample of the products of weathering derived from the catchment area upstream from the sampling point. Subject to modification by natural processes operating in the surface environment, the patterns of metal distribution in the bedrock and/or soil are reflected to varying degrees in the stream sediment.

By analyzing the trace-element content of sediment samples collected at densities ranging from 1 sample per 0.25 to 100 sq mi (commonly 1 sample per sq mi), it is often possible to detect anomalous patterns in the distribution of the ore metals related to mineral deposits (Hawkes and others, 1956; Hawkes and Webb, 1962; Boyle and others, 1966). This method of prospecting is now standard practice in many parts of the world. More recently, we have shown that by analyzing the same type of sample for a wide range of elements, including those involved in nutrition, it is also possible to detect geochemical patterns related to the geology as a whole and to the incidence of certain trace-element disorders in crops and livestock (Webb, 1964, 1965; Webb and others, 1964, 1965, 1968).

AGRICULTURAL APPLICATIONS

Trial surveys totalling more than 15,000 sq mi in the British Isles have yielded positive correlations between the composition of stream sediment and the incidence of cobalt deficiency in sheep (Webb, 1964), selenium toxicity and molybdenum-induced copper deficiency in cattle (Webb and others, 1964, 1965; Thornton, 1968), arsenic and lead toxicity in animals, copper and manganese deficiency in seedling spruce, and manganese deficiency in cereals (Thornton, 1968).

The following example from central England (Webb and others, 1968) is selected to illustrate the agricultural implications of stream-sediment reconnaissance surveys.

The geology of the study area, in the Southern Pennines of Derbyshire and North Staffordshire, is shown in Figure 1. Stream sediments were collected from tributary drainage at a mean density of 1 sample per square mile over all rock formations except the main limestone dome where there is no adequate surface drainage system. The samples were analyzed for 17 elements (Nichol and Henderson-Hamilton, 1965). Anomalous concentrations of 4 to 40 ppm of molybdenum were detected in streams draining parts of the lower Namurian shales east of the persistent cover of Newer Glacial Drift, compared with <4 ppm Mo elsewhere (Fig. 2). Similar stream sediment patterns also associated with the outcrop of these shales were found for arsenic, nickel, vanadium, chromium, cobalt, titanium, and iron. Analysis of rocks, soil, and pasture herbage shows that (1) marine facies of the Namurian shale are the principal source of the molybdenum, (2) residual soils derived from this rock are molybdeniferous, and (3) the pasture herbage growing on these soils tends to contain the higher concentrations of molybdenum (Table 1). Not all soils and pasture within the stream-sediment anomaly are molybdeniferous; both are normal over the interbedded sandstones and where the shales are concealed beneath outlying patches of exotic glacial drift.

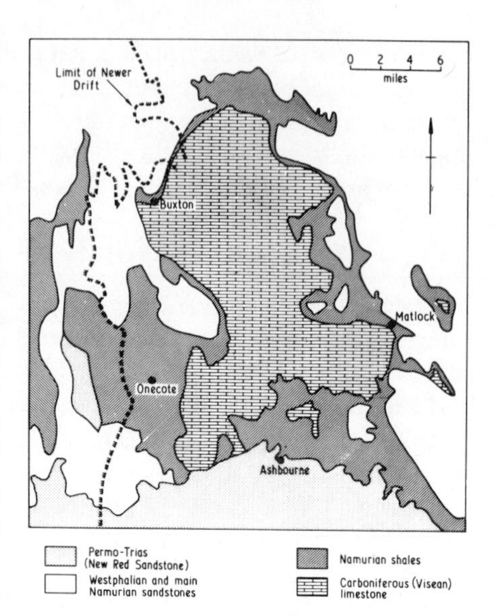

Figure 1. Simplified geology of the area covered by stream-sediment reconnaissance in central England (*after* Institute of Geological Sciences and B. K. Holdsworth, personal commun. Reproduced by courtesy of Nature).

The reported incidence of molybdenum-induced copper deficiency in cattle prior to the geochemical survey is given in Figure 3. Subsequent blood-copper tests carried out on 26 herds, excluding farms where hypocuprosis had previously been reported, showed the disease to be twice as prevalent within the stream-sediment anomaly compared to surrounding areas (Table 1; Fig. 4). It is important that the outward symptoms of copper deficiency were absent in several herds, despite low blood-copper values. As a further test, supplementation trials have since been carried out over the grazing season on farms within the anomalous area. Copper injections given to young stock grazing molybdeniferous pasture have yielded mean increases in live-weight gain of the order of 50 percent better than that of the control animals.

The magnitude of this particular problem in the United Kingdom, and the ease with which it can be assessed by stream-sediment reconnaissance are illustrated by the results of a survey covering 1000 sq mi of the Lias formation. The survey disclosed molybdenum anomalies totalling nearly 300 square

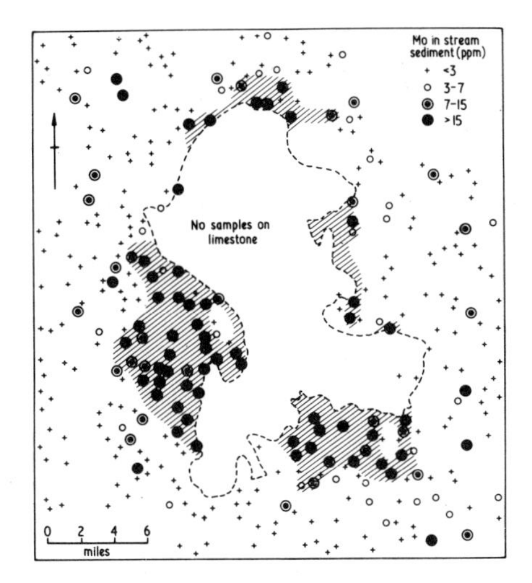

Figure 2. Distribution of molybdenum in the stream sediment of tributary drainage (reproduced by courtesy of Nature).

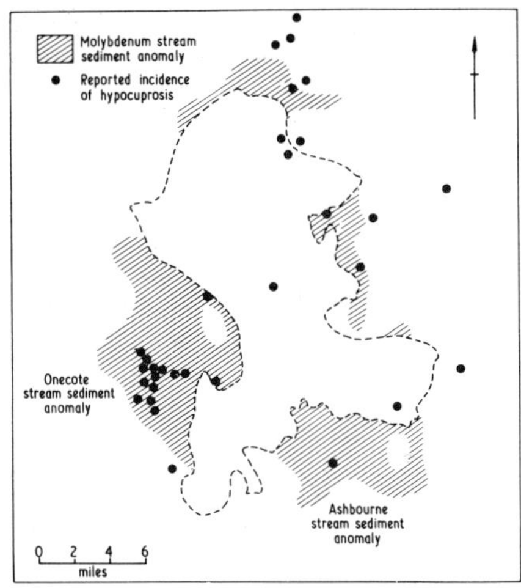

Figure 3. Incidence of hypocuprosis in cattle as reported before the regional geochemical survey. (Information from National Agricultural Advisory Service, the Veterinary Investigation Service and private veterinary surgeons. Reproduced by courtesy of Nature.)

miles wherein follow-up studies have confirmed the presence of molybdeniferous soils (*see* Fig. 5, which, incidentally, is a good example of significant metal variation within a single geological formation). The interim results of similar reconnaissance surveys now in progress show that suspect molybdeniferous areas exist over parts of several other geological formations and that the national total may well exceed 1,000,000 acres. Not all land within the anomalous areas is necessarily toxic, but the reconnaissance data, which are obtained with minimal effort, effectively define those areas meriting detailed study by more intensive methods of investigation.

The interpretation of stream-sediment data is by no means always as straightforward as might appear from the foregoing case history. Experience has clearly demonstrated that the chemical interrelations between stream sediments and the rock-soil-plant regime are varied, complex, and subject to the control of a bewildering array of processes and factors. Thus the relation between the composition of the sediment and the soil from which it is derived is liable to modification during the course of mechanical erosion, transport, and deposition along the stream channel and by differential leaching and precipitation from the stream waters, which are themselves subject to extraneous additions from the influx of circulating subsurface waters and rainfall.

The composition of the soil is dependent, not only upon the nature of the bedrock, but also upon the origin of the parent material, the nature and intensity of weathering and soil formation, the mobility of the elements, climate, topography, and other features of the environment. The relation between bedrock geology and the composition of ground and surface waters is

TABLE 1. RELATION BETWEEN THE INCIDENCE OF COPPER DEFICIENCY IN CATTLE AND THE MOLYBDENUM CONTENT OF STREAM SEDIMENT, ROCKS, SOIL, AND PASTURE HERBAGE, CENTRAL ENGLAND

	Mo (ppm)	
Sample type	Anomalous	Background
Stream sediment	4–40	<4
Bedrock		
Marine shale (73)	11 <2–50	
Nonmarine shale (11)		<2 <2–8
Sandstone (16)		<2 <2–5
Limestone (5)		<2
Residual topsoil (26)	16 4–35	2 <2–3.5
Grass (44)	3.4 1.0–8.0	2.0 1.6–5.0
Clover (21)	7.2 1.8–30.0	3.6 0.8–8.6
Percentage of copper-deficient cattle (280 animals tested)	77%	37%

Data refer to the range and mean metal content in the whole rock, in the -80-mesh fraction of soil and stream sediment, and in oven-dried plant material. Number of samples in parentheses.

likewise dependent upon these factors, and particularly on the *p*H and Eh of the environment.

The metal uptake of plants is also influenced by a great variety of factors, including the available metal content of the soil (as distinct from its total content), the relative proportions of other elements, the species and maturity of the plant, and climatic and seasonal variations. Other variations may arise from the use of fertilizers, soil amendments, and pesticides, and from atmospheric pollution.

Because of the complexity of these relationships it is hardly surprising that it is usually impossible to interpret stream-sediment data directly in terms of the precise chemical composition of the rocks, soil, and vegetation. Nevertheless, through knowledge of metal-dispersion characteristics in relation to specific environmental conditions, the development of diagnostic criteria and the use of statistical-analytical techniques, it is proving possible to predict probable relative variations in the composition of soils and vegetation from the patterns of metal distribution in stream sediments.

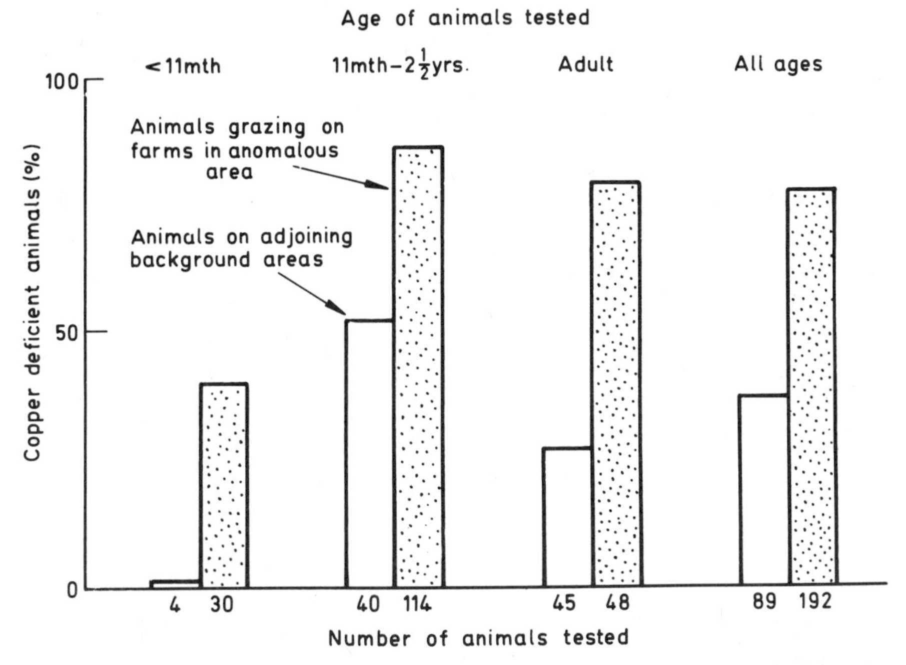

Figure 4. Incidence of copper deficiency (as indicated by blood copper analysis) in cattle, grouped according to age and the molybdenum content of the local stream sediment.

For the present, stream-sediment reconnaissance is necessarily a specialist operation, containing a strong element of research. The fact remains, however, that results to date confirm that, in many parts of the world, this technique can yield information bearing on the regional distribution of the elements and that this information can have a practical value, not only in mineral exploration but also in the context of health in agriculture.

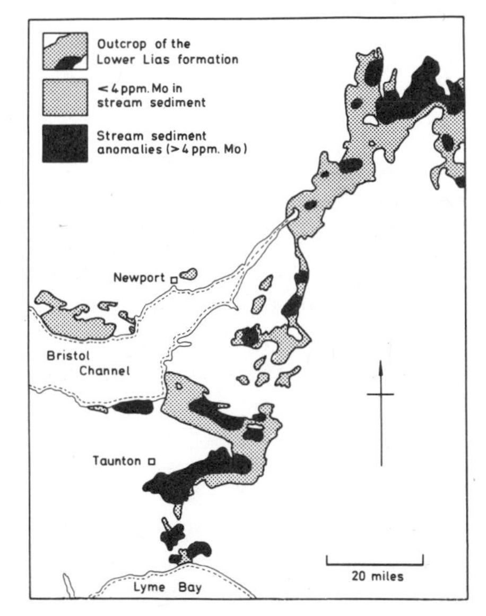

Figure 5. Distribution of anomalous molybdenum in stream sediment derived from rocks of the lower Lias formation in central and south-western England.

GEOMEDICAL IMPLICATIONS

The potential scope for geochemical reconnaissance in relation to health in man, although equally intriguing, has yet to be established. It is clearly difficult to envisage relations between geology and disease in highly urban communities with moving populations, varied diets, and bulked and processed foodstuffs. Nevertheless, even in these circumstances, the well-documented unexplained correlation between the composition of town water supplies and cardiovascular mortality (Schroeder, 1960, 1966; Morris and others, 1961; Crawford and others, 1968; Biorck and others, 1965) could conceivably have some geological connotations.

Geology may be more significant in epidemiology in less well developed territories where local communities, comprising a large part of the present world population, are still living in essentially primitive agricultural environments. Even in some developed countries, appreciable proportions of the population live relatively "close to ground" in the more rural areas. Neither is it uncommon for the smaller townships to be supplied with a significant proportion of food produced locally. In this connection, there is the correlation between the prevalence of dental caries and the molybdenum content of vegetables grown in market gardens supplying the towns of Napier and Hastings in New Zealand (Ludwig and others, 1960). There is also the unexplained correlation between the zinc-copper ratio in garden soils and certain cancers reported in North Wales (Stocks and Davies, 1964). National statistics for the United Kingdom given in Table 2 indicate that 10 to 25 percent of the common vegetables consumed by the total population are home-grown; these percentages must be much higher if only the rural population is considered. The extent to which sectors of developed societies can be dependent upon local foods is further illustrated by the fact that 3.25 million people in Britain drink milk supplied by local producer-retailers, and in some parts of the country the percentage of the local population so supplied exceeds 40 percent (Table 3).

Given comparable environments, there is often a measure of correlation between the metal content of the vegetation, the supporting soil, and the rock from which the soil was derived. Furthermore, these variations can differ from the norm by an order of magnitude or more. In general terms, stream-sediment reconnaissance can be expected to disclose patterns of variation that differ by factors of two or greater. Taken in conjunction with results obtained in the agricultural field, it seems reasonable, therefore, to believe that, in certain circumstances, stream-sediment patterns could reflect trace-element pat-

TABLE 2. NATIONAL STATISTICS FOR COMMON VEGETABLES PURCHASED AND CONSUMED (ENGLAND AND WALES, 1963)

	Consumed (oz./head/wk.)	Purchased (oz./head/wk.)	Percentage home-grown?
Potatoes	56.95	50.49	11
Cabbages	4.49	3.36	25
Sprouts	1.80	1.46	19
Leafy salads	1.23	0.93	25

terns in the human environment, particularly in rural areas and in the developing territories where a measure of correlation can be expected to exist between the geochemistry and the trace-element dietary intake of the local population.

In any given case, the epidemiological significance of the metal distribution patterns obtained by multi-element stream-sediment reconnaissance is a matter for interdisciplinary research. The geochemical data may prove to be irrelevant or present only a fraction of the evidence bearing on many epidemiological problems. In other instances, the geological-geochemical factor may assume greater importance. Quite apart from clinical disease, there is also the possibility that trace-element intake may be a contributory factor leading to substandard levels of general health. In developed societies, the possibilities of adverse imbalances arising from natural excesses are probably greater than imbalance due to natural deficiencies, although the latter could conceivably augment artificial deficiencies of some elements resulting from the process of refining certain staple foodstuffs such as flour and sugar. These and many other aspects of the problem are, for the present, necessarily conjectural; the pressing immediate need is for relevant objective facts.

In my view, the onus is on the geologist-geochemist to demonstrate with irrefutable case-history data that he can usefully contribute toward the solution of epidemiological problems. Only when we have done this may we legitimately and usefully enter the field of epidemiological controversy. We can, however, expect the constructive assistance of medical workers in carefully selected joint investigations of the geomedical premise. It would, for instance, be worthwhile to examine the significance of geochemical reconnaissance in relation to the previously mentioned examples of water supplies and cardiovascular disease, copper-zinc ratio and certain cancers, and molybdenum and dental caries. (Preliminary studies of the latter are now being carried out in conjunction with the Dental School, Birmingham University, in areas selected on the geochemical data shown in Figure 5.) Given adequate medical statis-

TABLE 3. PERCENTAGES OF TOTAL AND RURAL POPULATIONS SUPPLIED WITH MILK FROM PRODUCER-RETAILERS (NATIONAL STATISTICS FOR 1963)

Area	Population supplied with milk by producer-retailers	Percentage of total population served by producer-retailers	Percentage of rural population served by producer-retailers
England and Wales	2,741,000	5.7	29.8
NW Region	1,100,000	8.3	68.3
W. Region	187,000	16.0	45.6
N. Wales	127,000	21.8	43.9
Scotland	578,000	12.1	24.4
Berwick	13,000	58.5	58.5
Peebles	12,000	85.8	85.8

tics, it would also be interesting to examine regional stream-sediment data in relation to the incidence of several other diseases, including nephropathy in the Balkans (CIBA Foundation Study Group, 1967) and the controversial correlation of lead with multiple sclerosis (Warren and others, 1967). Quite apart from other considerations, the technique results in maps showing the distribution of a wide range of elements permitting the study of possible complex metal interrelations—it would be surprising if there were not metal interactions in man analogous to manganese and iron in plants and copper, molybdenum, and sulfate in animals.

On present evidence, it would be unwise to ignore the interdisciplinary potential of stream-sediment reconnaissance in medical geography. This plea is reinforced by the fact that vast areas are being sampled in many parts of the world for mineral exploration. There will undoubtedly be a growing coverage for agricultural purposes. It would require little additional effort to include a study of the geomedical implications. In any event, the economics of stream-sediment reconnaissance are such that the additional cost would be but a fraction of the outlay involved in a comprehensive epidemiological survey.

ORGANIZATION AND COST OF REGIONAL STREAM-SEDIMENT RECONNAISSANCE

The following estimates are based on two recent trial surveys covering 10,000 sq mi in the British Isles, including a territorial geochemical reconnaissance of Northern Ireland. Operation is geared to a unit rate equivalent to 20,000 sq mi per team-year.

The mean sample density is 1 sample per sq mi and accessibility is such that 90 percent of the samples can be collected at road-stream intersections. The unground −80-mesh fraction of the stream sediment is analyzed for 30 elements using an automatic spectrograph with direct typewriter and punch-card output. Plotting is done automatically by computer techniques supplemented by a Stromberg-Carlsen 4020 plotter (Nichol and others, 1966). Preliminary interpretation includes field checking of the principal geochemical patterns and statistical analysis.

Personnel requirement includes three geologist-geochemists, three analysts, and two data processors. Sampling is carried out by part-time labor operating in pairs from 1/4-ton vans. Each pair covers 200 sq mi per week. The three analysts, working an 8-hour day, handle 20,000 samples per year, yielding 600,000 items of data.

Approximate annual costs, exclusive of capital equipment, overhead, and computer time are:

Field and analytical staff	£13,000
Part-time labor (sampling)	5,500
Travel and field subsistence	5,000
Materials (analytical consumables, sample containers, and so on)	3,000
Preparation of provisional maps for selected elements	2,500
Contingencies	1,000
	£30,000

These costs, giving a unit rate of £1.5 per sq mi, reflect the benefits of working within the framework of the group's research program at Imperial College. Even so, given optimum gearing of staff, sampling rate, and analytical productivity, it should generally be possible to organize geochemical reconnaissance coverage at £2 to £3 per sq mi (based on current United Kingdom prices). These figures refer only to accessible terrain; sampling and field-checking costs rise steeply as vehicle-accessibility decreases.

CONCLUSION

By virtue of speed and low cost, stream-sediment sampling is ideally suited to geochemical reconnaissance on the regional scale. Results to date have demonstrated a useful range of application in agriculture and mineral exploration. The significance of stream-sediment data in medical geography is a matter for interdisciplinary research. Nevertheless, there are grounds for believing that a geochemical atlas could provide a unique source of ancillary information in epidemiological surveys, particularly in rural and emergent territories. Even in relatively urban areas, stream-sediment reconnaissance could provide data concerning the distribution of inorganic pollutants. Not least in importance are the possibilities of discovering new correlations between the elements and their interrelations when multi-element distribution maps are available for study alongside those showing geology, geography, and incidence of disease.

In view of the multipurpose potential of stream-sediment reconnaissance surveys, it would surely be logical to consider the advantages to be gained by collaboration in such surveys, at both national and international levels, of appropriate organizations concerned with mineral exploration, agriculture, and public health.

ACKNOWLEDGMENTS

Agricultural aspects of our research have been assisted by substantial grants from the Natural Environment Research Council. The geochemical reconnaissance of the area in central England was financed by the Institute of Geological Sciences, London. Animal tests were carried out in collaboration with the National Agricultural Advisory Service and the Veterinary Investigation Service.

REFERENCES CITED

Biorck, G., Bostrom, H. and Widstrom, A., 1965, On the relationship between water hardness and death rates in cardiovascular diseases: Acta Med. Scandinavia, v. 178, p. 239–252.

Boyle, R. W., Tupper, W. M., Lynch, J., Friedrich, G., Ziauddin, M., Shafiqullah, M., Carter, M., and Bygrave, K., 1966, Geochemistry of Pb, Zn, Cu, As, Sb, Mo, Sn, W, Ag, Ni, Co, Cr, Ba, and Mn in the waters and stream sediments of the Bathurst-Jacquet River District, New Brunswick: Canada Geol. Survey, Paper 65–42.

CIBA Foundation Study Group, 1967, The Balkan Nephropathy: London, Churchill.

Crawford, M. D., Gardner, M. J., and Morris, I. N., 1968, Mortality and hardness of water supplies: Lancet, v. 1, p. 827–831.

Hawkes, H. E., Bloom, H., Riddell, J. E., and Webb, J. S., 1956, Geochemical reconnaissance in eastern Canada, *in* Symposium on Geochemical Exploration: Internat. Geol. Congr. Trans., XX Session.

Hawkes, H. E., and Webb, J. S., 1962, Geochemistry in mineral exploration: New York, Harper & Row, 415 p.

Ludwig, T. G., Healy, W. B., and Losee, F. L., 1960, An association between dental caries and certain soil conditions in New Zealand: Nature, v. 186, p. 695–696.

Morris, J., Crawford, M. D., and Healy, J. A., 1961, Hardness of local water supplies and mortality from cardiovascular disease: Lancet, v. 1, p. 860–862.

Nichol, I., Garrett, R. G., and Webb, J. S., 1966, Automatic data plotting and mathematical and statistical interpretation of geochemical data, p. 195–210 *in* Proceedings of Symposium on Geochemical Prospecting in Ottawa: Geol. Survey Canada Paper 66–54.

Nichol, I., and Henderson-Hamilton, J. C., 1965, A rapid quantitative spectrographic method for the analysis of rocks, soils and stream sediments: Inst. Mining and Metallurgy Trans., v. 74, p. 955–961.

Schroeder, H. A., 1960, Relation between mortality from cardiovascular disease and treated water supplies: Jour. Am. Med. Assoc., v. 172, p. 1902–1908.

—— 1966, Municipal drinking water and cardiovascular death rates: Jour. Am. Med. Assoc., v. 195, p. 81–85.

Stocks, P., and Davies, R. I., 1964, Zinc and copper content of soils associated with the incidence of cancer of the stomach and other organs: British Jour. Cancer, v. 18, p. 14–24.

Thornton, I., 1968, The application of regional geochemical reconnaissance to agricultural problems: London Univ., Ph.D. dissert., 363 p. (available on interlibrary loan).

Warren, H. V., Delavault, R. E., and Cross, C. H., 1967, Possible correlations between geology and some disease patterns: New York Acad. Sci. Ann., v. 136, p. 657–710.

Webb, J. S., 1964, Geochemistry and life: New Sci., v. 23, p. 504–507.

—— 1965, Applied geochemistry and the community: Akad. Nauk. USSR. Geochem. Anal. Inst. Vernadsky Volume, p. 578–588.

Webb, J. S., and Atkinson, W. J., 1965, Regional geochemical reconnaissance applied to some agricultural problems in Co. Limerick, Eire: Nature, v. 208, p. 1056–1059.

Webb, J. S., Fortescue, J., Nichol, I., and Tooms, J. S., 1964, Regional geochemical reconnaissance in the Namwala Concession area, Zambia: Geochemical Prospecting Research Centre Tech. Comm. no. 47, 42 p.

Webb, J. S., Thornton, I., and Fletcher, W. K., 1968, Geochemical reconnaissance and hypocuprosis: Nature, v. 217, p. 1010–1012.

MANUSCRIPT RECEIVED BY THE SOCIETY JUNE 13, 1969

Printed in U.S.A.

THE GEOLOGICAL SOCIETY OF AMERICA, INC.
MEMOIR 123, 1971

Minor Elements in Water

MARVIN W. SKOUGSTAD
U.S. Geological Survey, Denver, Colorado

ABSTRACT

A complete characterization of natural waters includes a determination of the concentrations of 30 or more minor elements. Emission spectrochemical methods are uniquely suited to the determination of a comparatively large number of minor elements, mainly those classified as heavy metals. Spectrochemical data published by Braidech and Emory in 1935 and more recent data by Durfor and Becker, and by Barnett, Skougstad, and Miller, provide some insight into the minor-element character of the raw- and finished-water supplies of most of the major U.S. cities. Such data permit assessment of those supplies that contain either comparatively high or low concentrations of minor elements. The concentrations of many minor elements are unchanged by the municipal treatment-plant operations between the raw- and finished-water supply. On the other hand, part of the copper in a raw-water supply may be removed during treatment, even though the over-all result is a net increase in copper in the finished water. The aluminum concentration of finished water is greater than in the raw-water supply when aluminum sulfate is used as a coagulant aid. Significant amounts of iron may be removed during treatment.

CONTENTS

MINOR ELEMENTS IN WATER

Water is one of the simplest and, at the same time, one of the most complex of chemical substances. The chemical compound H_2O, water, is a very simple compound—just two atoms of hydrogen and one of oxygen. The physical and chemical properties of water have been studied more than those of any other chemical compound. As a result, we know a great deal about water and its behavior.

The term "water," however, is widely known in a broader sense, as, for example, lake water, ground water, sea water, drinking water, and so forth. In this sense, we are not dealing with water as a compound made up of a combination of two gaseous elements, but rather as an exceedingly complex system in which the chemical substance H_2O is the host medium for an extensive variety of both dissolved and suspended materials. Indeed, the unique properties of the chemical compound H_2O make it singularly suitable to serve as a host substrate in natural, geologic, and biologic processes. In fact, it may probably be said that the evolution of life as we know it, and life as we see it, has come about as a direct result of the chemical and physical nature of the chemical compound, water. As a result, it should be no surprise to us to know that human life may expire in so short a time as a few hours if wholly deprived of water. Other animal life, and plants as well, very quickly show the effects of a complete or even partial lack of water.

Water is called the universal solvent. There is scarcely any substance which will not dissolve in water to some measurable extent, given proper circumstances and conditions. Sometimes these circumstances may require an inordinately long period of contact time. On the other hand, high-temperature conditions or conditions of extremely high pressures may facilitate the solution of materials which might otherwise be thought of as insoluble. In such a manner, exceedingly diverse substances, as rocks, minerals, metals, and even complex organic substances, become a part of a natural water system. Obviously, a truly infinite variety of natural water systems is possible, depending on the source and the history of the water.

The complete characterization, therefore, of a natural water can be exceedingly complex. To some extent, the characterization can be segmented into reasonable units that have certain similarities with respect to either their chemical or physical nature, or with respect to the methods the scientist uses to detect or evaluate them. Thus, we might classify our examination of a natural water according to the following scheme:

Category	Characteristics	Examples
1.	Major solutes and properties	Hardness, salinity, Ca, SO_4, and others
2.	Minor elements	Mo, Pb, V, Cr, Ni
3.	Radio-isotopes	U, I^{131}, Cs^{129}, Ra
4.	Organic solutes	Detergents, pesticides
5.	Suspended particulate matter	Turbidity, clays
6.	Microbiological	Bacteria, rotifers, larvae, viruses, and others

In the classification above, the minor-elements category is shown in a proper relationship to the several other categories, all of which may or may not be considered in connection with an evaluation of a natural water. Actually, the popularity or even the importance of some of these categories varies from time to time—some, of course, more than others. For example, immediately following World War II when both the U.S.S.R. and the U.S. were conducting nuclear-weapons tests in the atmosphere, there was great concern about the hazard of radioactive fallout accumulation in water supplies, as well as in other materials we use. Since a moratorium was placed on atmospheric testing in 1963, the urgency of fallout protection measures has greatly diminished. Although the hazards of radioactive fallout are thus not of immediate concern, the water-quality specialist continues to monitor radioactivity which may occur from natural sources (background), from atmospheric detonation of nuclear devices, or from accidental release of isotopes.

Other examples of the rise and fall in favor of certain water-quality characteristics may be noted, particularly with regard to detergents, to pesticides, and more recently to "nutrients." The latter term may be ambiguous, however, for what may be nutritious to one form of life may be poison to another, and what may be nutritious at one level of concentration may be toxic at a higher level.

Minor elements, while generally of some curiosity, have never been popular as a subject of investigation in water-quality studies. This has been due to lack of information on the significance of minor-element concentrations and also on the cost of providing any significant amount of minor-element data. The paucity of data on concentration levels of minor elements is particularly apparent when the amount of such data is compared with that accumulating continuously on the major solutes.

The inorganic, nonradioactive chemical elements, radicals and compounds that occur naturally, or are added during water-treatment practices, or occur as a result of man's stress upon natural conditions, may then be classified as either major or minor, depending upon what might be observed to be the common or natural circumstance. For the purpose of defining major and minor elements it seems reasonable to assign a concentration level of 1 mg/l (milligram per liter) as an appropriate demarcation. Any solute which commonly occurs at concentration levels greater than this is a major constituent and includes such solutes as sodium, potassium, calcium, magnesium, bicarbonate and carbonate, chloride, sulfate, and, sometimes, nitrate and phosphate and even silica. The minor elements, then, are those whose concentration levels do not ordinarily exceed 1 mg/l and include not only common inorganic solutes such as iron, manganese, and aluminum, but also the more unusual solutes—the less common alkali and alkaline-earth metal ions, the whole group of transition-series elements, those of the inner transition series, the rare earths, and a number of elements in Groups IIIA through VIIA of the Periodic Table, of which some representatives are boron, gallium, tin, arsenic, selenium, and bromine and iodine.

Of this rather extensive list of minor elements, quite a few can be eliminated from further consideration for either of two reasons: the element itself is so rare that the chances of its occurrence in water at a significant level of concentration is practically zero, or the convenient and common analytical methods are of inadequate sensitivity to detect the extremely small amounts that may be present. Examples of the first case include most of the rare earths. Although methods of adequate sensitivity exist, the occurrence of these substances in natural water is so rare that they are almost never found. Occasionally, traces of cerium or ytterbium have been detected.

Some minor elements cannot easily be determined at the same low concentration levels as others, inasmuch as the sensitivity of the analytical technique varies greatly from element to element. Elements such as bismuth and antimony, selenium and tellurium are not always included in a routine minor-element examination because their determination is tedious and costly rather than because traces are difficult to detect. The analysis of a water sample for its minor-element content is always a compromise. Ideally, every minor element present at a detectable level of concentration should be determined and that determination should be made by the most sensitive, but, at the same time, the fastest and most economical method known. Such a practice could be very expensive but would ensure maximum useful data from every sample collected. To use the most sensitive analytical method for each element generally means that a separate sample aliquot be taken for each determination—which may suggest that an unusually large sample may be required in order to handle all the separate determinations. Obviously, data so acquired can be very costly because of the many individual, time-consuming, analytical procedures that have to be followed.

An appropriate compromise, then, is to find a method or technique whereby several or many elements can be determined in a single sample aliquot. If necessary, two or three aliquots will provide sufficient sample for the determination of 20 or more elements. Unfortunately, it is inevitable that the method most suitable for multi-element determinations on a single aliquot is not the method of highest sensitivity for each of the minor elements of interest.

The one analytical technique which, so far, surpasses all others in the number of minor elements which may be quantitatively determined in a single sample aliquot is that based on emission spectroscopy. Fortunately, and with but few exceptions, emission spectrographic methods are sufficiently sensitive for a majority of the minor elements which are of principal interest. The notable exceptions are the non-metallic trace elements, bromine and iodine, arsenic, selenium and tellurium, and a few metallic elements including zinc, cadmium, and, perhaps, bismuth. The sensitivity of the emission spectrograph is satisfactory or at least acceptable for most of the other minor elements that are of interest in water. Those elements which can be conveniently determined are noted in Figure 1.

A number of emission spectrographic techniques have been devised for analyzing samples in aqueous solution. Few of these are suitable for analyzing environmental water samples, as the methods must have adequate sensitivity for the maximum number of elements. Direct analysis of a water sample without either prior concentration of the sample or separation and concentration of the minor elements is best carried out by either the porous- or vacuum-cup techniques, or the rotating-disk technique. A copper- or silver-spark technique has been used with some success. The latter, however, should be considered a means of preliminary separation inasmuch as the solvent is evaporated, leaving a residue of solutes on the electrode for excitation. In that respect, the technique is similar to the total-residue technique wherein a sample aliquot is chosen to yield approximately 100 mg of evaporated residue. An appropriate aliquot is evaporated to dryness and the resulting residue recovered, mixed until uniform in composition, and then further mixed with pure graphite and loaded into a cupped, graphite electrode for excitation in a DC arc. Increased

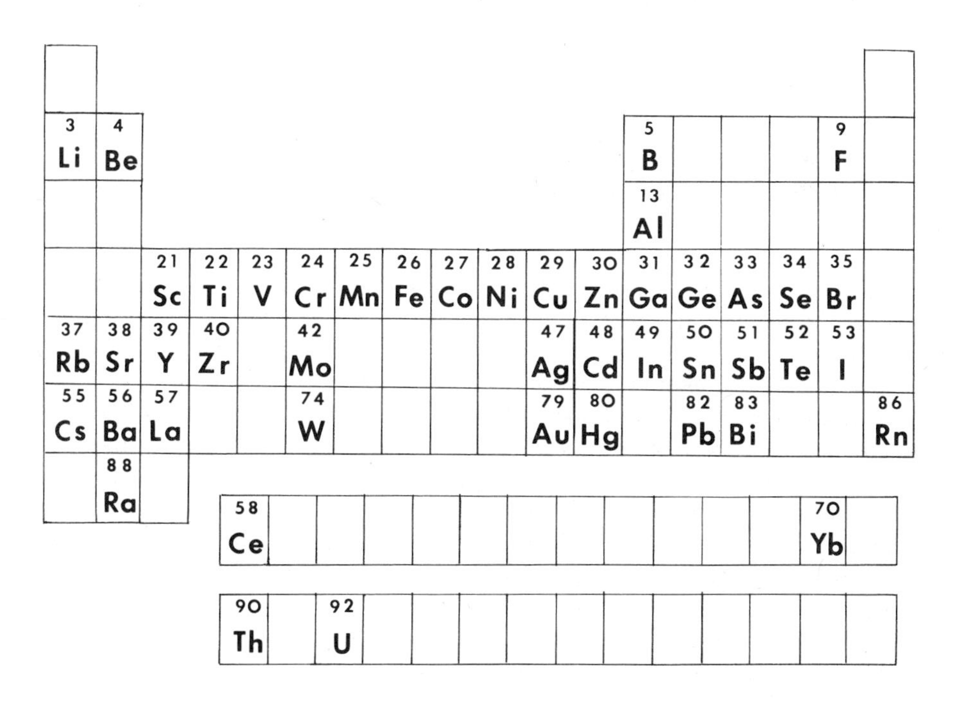

Figure 1. Important minor elements in water.

accuracy can be achieved by adding a selected buffer-matrix material, although such practice proportionally reduces the sensitivity of the method. When a slight sacrifice in analytical sensitivity can be tolerated, the gain in reliability of the resulting final data more than justifies the practice.

The total residue method, although subject to some limitations, offers the best means of determining the largest number of minor elements in a sample, as all the constituents except those whose compounds volatilize during the evaporation of the sample will remain with the residue and will be excited when the sample is arced. An obvious limitation of this method applies to those samples whose concentrations of important minor elements is so low and whose concentrations of major solutes is so high that an aliquot of suitable size cannot be evaporated. The usual cut-off point for water is approximately 1000 mg of solutes per liter. Samples containing more than this amount require such a small volume to provide the necessary 100 mg of residue that severe lowering of the detection limit occurs.

The limit in the other direction, of course, is restricted only by the patience of the analyst in evaporating the large volumes of water needed to obtain sufficient residue for analysis. Because the detection limit becomes poorer with increasing concentrations of dissolved solutes, the total-residue technique is applicable only to fresh water; that is, water containing less than about 1000 mg/l of solutes.

When slightly saline or more highly mineralized samples are to be analyzed, a separation of the minor elements from at least the bulk of the major solutes is desirable. Several techniques are useful, including both ion exchange and precipitation; the latter is usually the more convenient. As pointed out previously, the most serious limitation of such separation techniques is the comparatively small number of minor elements that can be collected from a single aliquot. For example, precipitation systems are quite inadequate for the rare alkali and rare alkaline-earth metals, nor are all of the transition metals precipitated with equal efficiency. Such methods, however, do have several offsetting advantages. The resulting matrix, which consists of the added carrier precipitate, can be chosen with care; furthermore, it is identical in composition for all the samples. Moreover, the method is equally applicable to both fresh and saline water samples.

In spite of the restriction of most other analytical methods to one or, at the most, a few elements which can be determined in a single sample, there are a number of methods which can be used when much greater sensitivity is required. Colorimetric and spectrophotometric methods are very sensitive and extensively used, but nearly always applicable to no more than one or two elements per sample aliquot. Obviously, the cost of determining more than a few minor elements by spectrophotometry is prohibitive. Fluorometric methods may offer greater sensitivity than spectrophotometric methods, but suffer from the same disadvantage. In addition, while reasonably good or excellent ultraviolet and visible spectrophotometers are available in nearly every laboratory, not all labs have good fluorometers.

Polarographic methods offer adequate sensitivity, but are not widely used, probably because of the difficulties in training technicians to operate the equipment on a completely routine basis.

SAMPLING

The manner in which a sample is collected is perhaps more important in the case of water analysis than in the analysis of most other materials. Water in nature is almost always a continuously changing material, so that a sample collected at one moment may have a different composition from one collected at the same place some time later. This is less true for ground water than surface water, particularly in sampling streams that exhibit rather wide fluctuations in composition as a result of intermittent influxes of waste discharges. Although the time of sampling may be of only slight importance when major constituents are determined, it is of considerable importance with respect to minor elements. In this connection, practices which are unquestioned when analyzing water for its principal constituents may be completely unacceptable when minor elements are to be measured.

For this reason, any spot checking of a water supply should be questioned. Does the sample represent an average composition, or could it represent an isolated extreme of short duration? Ideally, water-analyses measurements should be based on a continuous record of sufficient duration to define the time scale over which significant compositional changes are likely to occur. Once this is established, the investigator can determine the sampling frequency required to detect significant changes.

A continuous record of total composition at a selected site is quite impossible, but experience has demonstrated the validity of certain correlations of measurements which can serve as a workable substitute. For example, it has been established that major variations in the composition of water in a large ground-water aquifer generally occur over a comparatively long period of time. Similarly, the chemical composition of the water in a large reservoir changes slowly with time, except locally, where the characteristics of the inflow may be identified before being mixed with the body of the reservoir. On the other hand, experience has shown that compositional variations of a stream may be large and may vary considerably over a relatively short period of time. The most important factor in stream composition is the rate of flow or discharge rate. At periods of high flow, for example in the spring and early summer, the concentrations of solutes are lowest, although the total solute load will generally be highest. In late summer, fall, and winter, the solute concentrations are ordinarily highest.

The importance of these factors with regard to sampling for minor-element analysis must be emphasized. Minor-element concentrations may or may not parallel the concentrations of the major ions. However, it is certain that fluctuations in concentrations do occur—and the risk of obtaining misleading data or, at least, atypical or nonrepresentative data, is too great to overlook.

Any analytical data based on a single, one-shot sample must be used with a certain amount of caution.

Sampling ground water is considerably less critical with respect to time than is the sampling of surface water. Nevertheless, certain precautions are necessary, and the sampler must assure himself that the sample he is collecting does, in fact, represent the conditions which he ascribes to it. At the very minimum, the well should be pumped for sufficient time to ensure that the character of the water brought to the surface is truly representative of the water in the aquifer. This is important if the well has not been in use for some time, and water has accumulated in the well casing.

The problems of sampling and analysis which must be considered when determining the minor-element character of water lead, then, to this query. Assuming that the analytical techniques provide minor-element data of satisfactory reliability, does the sample then represent the conditions about which we need information; and has the manner of collecting, handling, and storing the sample altered it in any way?

MINOR-ELEMENT CONCENTRATIONS

Having considered briefly the role of water as a solvent system, and having examined some of the analytical and sampling problems associated with the determination of minor elements, let us now look at some of the findings.

One of the earliest extensive reports on minor-element concentrations in public water supplies was published by Braidech and Emory (1935). Their analyses, based on a semiquantitative, emission spectrographic method, included data for 19 minor elements in the supplies of 24 major United States cities. Later analyses of water from some of these same sources have indicated a few apparent errors in the Braidech and Emory data, so that not all of their work can be considered as a reliable index of the concentration levels of trace elements in these water supplies. For example, the strontium values given by Braidech and Emory are generally too high by a factor of about 10. Likewise, many of their values for silver appear high by a factor of at least 10, by comparison with data obtained more recently.

The next comprehensive assemblage of data on minor elements in public water supplies was not published until nearly 30 years later. Durfor and Becker (1964), in an evaluation of the supplies of the 100 largest cities in the United States, included chemical data on the major cations and anions, and also spectrographic minor-element and radiochemical data. Their analyses are especially helpful because they include a great many samples from both raw- and finished-water supplies. A regrettable shortcoming in the Durfor and Becker data is that only a single analysis was made of one collection in each case. The variations and ranges of minor-element concentrations over a period of several months for each of these cities would be a valuable contribution to our store of knowledge.

An examination of the Durfor and Becker data reveals some interesting facts concerning minor elements in public water supplies. First, consider the maximum and minimum concentrations of several minor elements (Table 1). These indicate the range of concentration for several of the most important minor elements. Molybdenum and vanadium concentrations as high as 68 and 70 μg/l (micrograms per liter), respectively, were encountered. A later study in 1969 by Barnett, Skougstad, and Miller (*see* Table 2) has demonstrated the occurrence of molybdenum concentrations of up to 240 μg/l in the supply of a major U.S. city resulting from a recent change in source (Table 2). Lead concentrations as high as 62 μg/l were found by Durfor and Becker, and concentrations of titanium, chromium, and nickel of 49, 35, and 34 μg/l, respectively, were found. The maximum amount of silver found in any supply was only 7.0 μg/l, which contrasts with a maximum of 200 μg/l reported by Braidech and Emory in 1935. Braidech and Emory reported no city supply having a silver concentration of less than 10 μg/l. Furthermore, the Denver supply, in which they reported 200 μg/l of silver, was found by Durfor and Becker, upon analysis of six separate samples, to show no detectable silver in four of the samples, and less than 0.26 μg/l in the other two samples. The Durfor and Becker data have been substantiated by Barnett, Skougstad, and Miller.

Maxima of more than 1000 μg/l are recorded for four elements: aluminum, iron, manganese, and strontium. The median strontium concentration of 110 μg/l may be surprising, although previously reported data on the occurrence of strontium in surface and ground water (Skougstad and Horr, 1963) indicate appreciable concentrations of strontium in many water sources. Of the 100 cities included in the Durfor and Becker report, 11 had strontium concen-

TABLE 1. MINOR ELEMENTS IN PUBLIC SUPPLY, 1962*
(Finished water—100 U.S. cities)

		Maximum μg/l	Median μg/l	Minimum μg/l
Aluminum	Al	1500	54	3.3
Barium	Ba	380	43	1.7
Boron	B	590	31	2.5
Chromium	Cr	35	.43	—
Copper	Cu	250	8.3	—
Iron	Fe	1700	43	1.9
Lead	Pb	62	3.7	—
Lithium	Li	170	2.0	—
Manganese	Mn	1100	5.0	—
Molybdenum	Mo	68	1.4	—
Nickel	Ni	34	—	—
Rubidium	Rb	67	1.0	—
Silver	Ag	7	.23	—
Strontium	Sr	1200	110	2.2
Titanium	Ti	49	—	—
Vanadium	V	70	—	—

*Adapted from Durfor and Becker (1964)

trations of less than 6.0 μg/1, and 6 had concentrations exceeding 700 μg/1. Three cities supply water containing more than 1000 μg/1 of strontium: San Diego, California, and El Paso and Lubbock, Texas (Table 3).

In Durfor and Becker's minor-element data on the 100 largest cities, many of the minor elements were undetected in the supply of one or more of the cities included in the study. Because of some variation in the detection limits for these elements, the calculation of a median concentration may be futile. Of perhaps greater significance is the distinction between cities having comparatively high concentrations of a given minor element and those having comparatively low concentrations. Four minor elements were found at maximum concentrations of intermediate level, that is, of the order of a few hundred μg/1. These elements (Table 4) are barium, boron, copper, and lithium. Both barium and boron were found in all samples tested. The median concentrations of these two elements were 43 and 31 μg/1, respectively. Of particular interest is the comparatively high concentration of barium, 380 μg/1, found in one supply.

If we look at the cities with comparatively high and low concentrations of barium (Table 5), we find 11 cities whose public supplies contain no more than 10 μg of barium per liter, and 6 cities in whose supplies the barium concentration is 200 μg/1 or more. The highest concentration found in any public supply was 380 μg/1 at Houston, Texas; the lowest, 1.7 μg/1, at Tacoma, Washington.

These data, although interesting,

TABLE 2. MOLYBDENUM CONCENTRATION, DENVER WATER SUPPLY, 1966*

Date	Kassler Treatment Plant Raw water μg/l	Kassler Treatment Plant Finished water μg/l
5/13	2	2
6/10	14	4
7/ 8	54	27
9/16	260	240

*Adapted from Barnett and others (1969)

TABLE 3. STRONTIUM IN PUBLIC WATER SUPPLIES, 1961*
(Median concentration, 110μg/l)

Less than 6 μg/l		More than 700 μg/l	
City	Concentration μg/l	City	Concentration μg/l
Birmingham	2.2	Tucson	850
Denver	4.1	San Diego	1100
Bridgeport	3.8	Charlottesville	840
Atlanta	2.4	Columbus	740
Savannah	2.0	El Paso (1)	760
Jersey City	5.3	(2)	1200
Charlotte (1)	2.4	(3)	1000
(2)	5.3	(4)	1000
Portland	2.9	Lubbock (1)	1200
Seattle	2.6	(2)	6300
Tacoma	2.6		

*Adapted from Durfor and Becker (1964)

fail to define concentration ranges and variations in minor-element concentrations in any one city over a period of time. That this can be important was demonstrated by Barnett and others (1969) in a study of the public supply of the City of Denver. They sampled both raw and finished water monthly for a period of 5 months in an attempt to define concentration ranges which might be expected for as many minor elements as could be detected. One of their unanticipated findings was a marked increase in the molybdenum concentration in one of the sources of supply for the Denver system (Table 2).

The explanation for this marked increase in molybdenum concentration was very obvious. One of the large reservoirs providing water to a portion of the Denver system was not tapped until the middle of June, so the first two monthly samples contained only very low concentrations of molybdenum. The

TABLE 4. MINOR ELEMENTS IN PUBLIC WATER SUPPLY, 1961*
(Finished water—100 U.S. cities)

		Maximum μg/l	Median μg/l	Minimum μg/l
Barium	Ba	380	43	1.7
Boron	B	590	31	2.5
Copper	Cu	250	8.3	—
Lithium	Li	170	2.0	—

*Adapted from Durfor and Becker (1964)

TABLE 5. BARIUM IN PUBLIC WATER SUPPLIES, 1961*
(Median concentration, 43 μg/l)

10 μg/l or less		200 μg/l or more	
City	Concentration μg/l	City	Concentration μg/l
Albany (1)	10	Albuquerque	210
(2)	9.5	El Paso	210
Birmingham	10	Houston (1)	340
Charlotte	8.9	(2)	380
Denver	6.2	Lubbock	210
Greensboro	8.6	Oklahoma City	260
Honolulu (1)	5.4	Rockford (Ill.)	200
(2)	4.6		
Miami	5.7		
Portland	4.4		
Seattle (1)	3.1		
(2)	5.7		
Springfield (Mass.)	8.1		
St. Petersburg	5.1		
Tacoma	1.7		
Worcester	8.1		

*Adapted from Durfor and Becker (1964)

TABLE 6. VARIATION IN CONCENTRATION OF MINOR ELEMENTS, DENVER PUBLIC SUPPLY (RAW WATER), 1966*

Date	Element concentration, μg/l				Dissolved solids
	Ba	Cu	Fe	Mn	μg/l
5/13	48	120	10	9	255
6/10	50	85	5	34	287
7/ 8	40	45	8	56	295
8/ 5	37	50	10	29	285
9/16	48	35	5	11	227

*Adapted from Barnett and others (1969)

increase in molybdenum concentrations, beginning with the samples collected in July 1966, reflect the increasing proportion of the water supply being provided by the one large reservoir and continuing the rest of the summer. One of the main streams feeding this particular reservoir drains an area of large-scale molybdenum mining operations. This results in a comparatively high concentration of molybdenum in the reservoir and in that portion of the city supply being provided water from this source.

TABLE 7. COPPER IN RAW AND FINISHED WATER AT THE DENVER NORTH MARSTON TREATMENT PLANT, 1966*

Date	Raw water μg/l	Finished water μg/l
5/13	120	20
6/10	85	12
7/ 8	45	8
8/ 5	50	7
9/16	35	6

*Adapted from Barnett and others (1969)

Other variations in minor-element content were noted in the various Denver supply sources (Table 6), but none quite as dramatic as those observed for molybdenum.

Treatment-plant operations may, in some cases, alter the minor-element content of the raw-water supply, either adding or removing minor elements. The study by Barnett and others (1969) amply demonstrated this. As an example, raw-water storage reservoirs are sometimes treated with copper sulfate to inhibit algal growth. One of the Denver city reservoirs was treated with copper sulfate in late March. The raw water in this reservoir showed 120 μg/l of copper when sampled in May, but only 35 μg/l when sampled 4 months later (Table 7). There was a steady decrease in copper concentration in the raw-water intake to this plant during this time. The finished water showed a similar decrease in copper content, although the amount of copper in the finished water was always much less than in the raw-water supply, indicating a removal of copper by the treatment operation.

Aluminum sulfate is sometimes added during treatment to aid in coagulation of suspended matter. This may result in an increase in aluminum concentration in the finished water over that originally present in the raw-water supply. The data of Barnett, Skougstad, and Miller demonstrate this. At the

North Marston Treatment Plant, the mean aluminum concentration in the raw water over a five-month period was 74 μg/l. The mean concentration of aluminum in the finished water over this same period was 180 μg/l, nearly 2.5 times the mean for the raw water.

There was no conclusive evidence to indicate any change in concentrations between raw and finished water for barium, boron, lithium, molybdenum, rubidium, strontium, and zinc. As has been mentioned before, at least a part of the iron may be removed by the treatment process.

SUMMARY

Environmental waters constitute complex chemical systems whose dissolved-solute compositions frequently include a great number of minor elements. By careful sampling and precise analysis the more significant of these minor elements can be determined, and a water sample rather completely characterized as to its minor-element content. Emission spectrochemical methods are the most useful for identifying and measuring the concentrations of an extensive number of minor elements in natural waters. Spectrochemical data by Braidech and Emory; Durfor and Becker; and Barnett, Skougstad, and Miller have served to characterize the water supplies, both raw and finished water, of the major cities in the United States. Certain minor elements that predominate in the waters of various sections of the country may be lacking in other sections. The treatment process employed to convert a raw-water supply to finished water suitable for public consumption may add minor elements to a water supply, as in the case of aluminum, or may remove them, as in the case of iron and copper.

REFERENCES CITED

Barnett, P. R., Skougstad, M. W., and Miller, K. J., 1969, Chemical characterization of a public water supply: Am. Water Works Assoc. Jour., v. 61, p. 61–67.

Braidech, M. M., and Emory, F. H., 1935, The spectrographic determination of chemical constituents in various water supplies of the United States: Am. Water Works Assoc. Jour., v. 27, p. 557–580.

Durfor, C. N., and Becker, E., 1964, Public water supplies of the 100 largest cities in the United States, 1962: U.S. Geol. Survey Water-Supply Paper 1812, 364 p.

Skougstad, M. W., and Horr, C. A., 1962, Chemistry of strontium in natural water: U.S. Geol. Survey Water-Supply Paper 1496, 97 p.

PUBLICATION AUTHORIZED BY THE DIRECTOR, U.S. GEOLOGICAL SURVEY, WASHINGTON, D.C.

MANUSCRIPT RECEIVED BY THE SOCIETY FEBRUARY 16, 1970

Printed in U.S.A.

THE GEOLOGICAL SOCIETY OF AMERICA, INC.
MEMOIR 123, 1971

Regional Plant Chemistry as a Reflection of Environment

J. F. HODGSON
AND
W. H. ALLAWAY
U.S. Plant, Soil, and Nutrition Laboratory, Agricultural Research Service, U.S. Department of Agriculture, Ithaca, New York

AND
R. B. LOCKMAN
Agrico Plant Analysis Laboratory, Continental Oil Company, Washington Court House, Ohio

ABSTRACT

The present period reflects a growing concern over the dietary levels of elements important to human health. Transportation of food from one area to another, greater variety of foods from which to choose, and introduction of certain food supplements are important in reducing elemental deficiencies and masking the effect of regional variations in elemental composition of crops on diets. Nonetheless, differences in amounts and availability of elements from one region to another appear to be reflected in the levels of an element in human or animal tissues or in the incidence of animal disease. Regional variations in Zn content of plant tissue are compiled from the literature and compared with corresponding values from a national sampling program for corn and alfalfa. They reflect low levels of Zn most commonly in the South, Great Plains, and the Southwest. These areas generally corresponded with locations of low Zn content of human blood.

CONTENTS

Table

This paper is concerned with the geographic distribution of certain elements in plants and how meaningful this distribution may be to the development of endemic diseases of man.

Human nutrition in the United States has improved steadily over the past half century as reflected in the physical stature of our citizens, the abilities of our athletes, and the acceptance rate of draftees for military service. Our improved diets can be traced to, among other things, a greater variety of foods from a wider geographic area. The liability of a diet derived from a small number of species grown in a given locality has been steadily reduced. The contribution of supplements such as iodine to salt and iron, calcium and many vitamins to selected foodstuffs has also been important.

The effect of regional variations in sources of food was clearly evident in 1916 when 1.5 to 3 percent of the army draftees from the Northern States were rejected because of endemic goiter (McClendon, 1939). The use of iodized salt now obscures variations in the natural supply of iodine to human diets, but it is likely that the probability of goiter is further reduced by the introduction of produce from coastal areas and, especially, seafood onto the shelves of inland supermarkets.

Iodine remains to this day the only element proven to give rise to an endemic deficiency in humans resulting from a shortage of the element in plants. McClendon (1939) comments that " . . . goiter maps and iodine maps, supposing they existed, would fit like the land fits the ocean, goiter incidence and iodine varying inversely."

Deficiencies of Fe and Ca are also traced to dietary sources, but in these cases it has not been possible to associate the deficiencies with geographic variations in the amount of these elements in food. This is noteworthy because the distribution or availability, or both, of each of these elements in soils varies greatly across the United States. Apparently, dietary selection, species variations, plant uptake, and other factors have too much influence over the supply of these elements in the human diet for geographic patterns of soil content or availability to be expressed in human deficiencies. These two elements are now used to enrich many foods. This reduces the incidence of deficiency in human diet and further masks any geographic pattern that might otherwise be evident. The factors that tend to obscure geographic distribution will be considered further in connection with Zn.

Fluoride deficiencies are known to show a geographic pattern, but the source of supply for humans is principally from water and not plants. This is due to the very limited uptake of F by most plants.

To the above limited number of human nutritional disorders that are known to relate to inorganic plant composition can be added a host of suspected ones. Deficiencies of Cr, Zn, Mo, Mn, and P, and toxicities of Pb, Cd, As, Cu, and Be are only a few of the potential sources of concern.

Farm animals are generally more dependent on locally produced crops than on man; thus in regard to ruminant animals, for example, geographic distribution of Co and Se deficiencies is easily traced to low levels of these elements in plants of certain areas. Maps illustrating these areas are reproduced in Figures 1 and 2 from Kubota (1968) and Kubota and others (1967). The deficiency diseases, white muscle disease of calves and lambs low in Se and Bush sickness of cattle low in Co, often follow the same distribution.

To consider what we can and cannot learn about the distribution and correction of nutritional problems from a knowledge of plant composition, the Zn content of crops will be examined. To begin with, we must recognize that there is only limited indication of Zn deficiency in humans and that within the United States, there is no geographic pattern presently associated with reported responses to Zn administration to humans. But in contrast with some elements to which we are increasingly being exposed, we are witnessing a gradual disappearance of many artificial sources of Zn. We are becoming increasingly isolated from galvanized articles and water pipes which provided a constant supply of Zn to former generations. This has placed increasing reliance on plants as our source of Zn, and there is doubt that this source is adequate for the need. Can we then anticipate patterns of Zn deficiencies in humans by looking into geographic variations in plant content of Zn?

The amount of Zn in plants varies with such diverse factors as soil content and availability, plant species and variety, season, temperature, light, stage of growth and, perhaps as important as any, a host of nutrient interactions. To give perspective, the dependency of Zn content of plants on geographic features will be examined and then compared with certain other sources of variation in the Zn content of plants.

Beeson (1959) has recorded those areas where Zn deficiencies had been observed in various crops prior to 1957. Zn deficiencies in many areas of the Great Plains, particularly in Colorado, North Dakota, and Nebraska, have since been found. Generally, calcareous soils and sandy soils low in Zn are most likely to produce Zn-deficient crops.

The distribution of Zn deficiencies compares favorably with areas of low Zn content in plants. The amount of Zn in forages and corn and a scattering of other species is given in Table 1 and Figure 3. Values in Table 1 are taken from the literature. The data in Figure 3 were accumulated by Lockman with the collaboration of L. G. Horrill. The latter information was gathered in

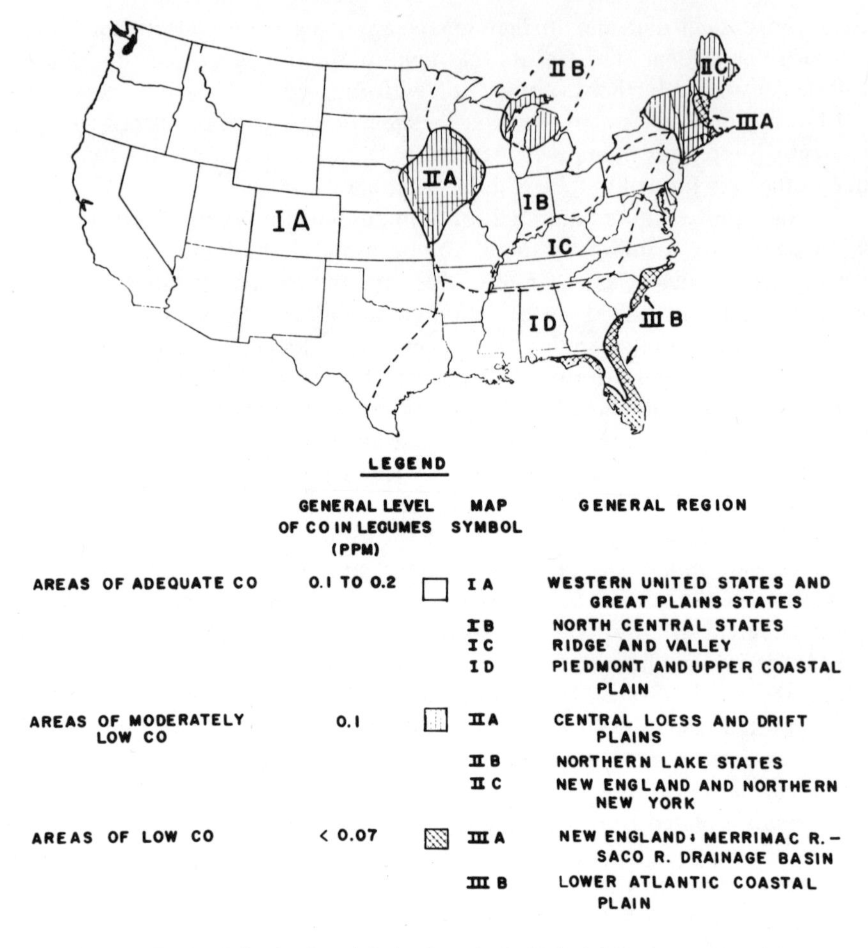

Figure 1. Regional distribution of Co in plants in the United States (ppm dry weight).

connection with a survey of plant growth problems of unknown origin and, therefore, does not reflect a totally representative sampling. All corn samples were ear leaf at bloom stage; alfalfa was upper third of plant at late vegetative through bloom stage. For comparison, contents of Cu, Fe, and Mn are given in Figure 4 for the alfalfa samples recorded in Figure 3.

From the results of French and others (1957) in Pennsylvania (Table 1), it is apparent that the Zn content of red clover is related to that of the parent rock. The correlation was highly significant in their statistical analysis, but over-all differences were not large. Regional variations in Zn content of alfalfa and corn can best be seen in Figure 3. The means from the relatively Zn-deficient regions of the Great Plains and Southeast range from 25 to 33 ppm Zn in oven-dry alfalfa and, excluding the Southern Piedmont and mountain sections, from 19 to 32 ppm in corn. Northeast and North Central regions have means ranging from 33 to 45 ppm Zn in alfalfa and from 28 to 68 ppm in corn. Differences, again, are certainly evident and, considering that

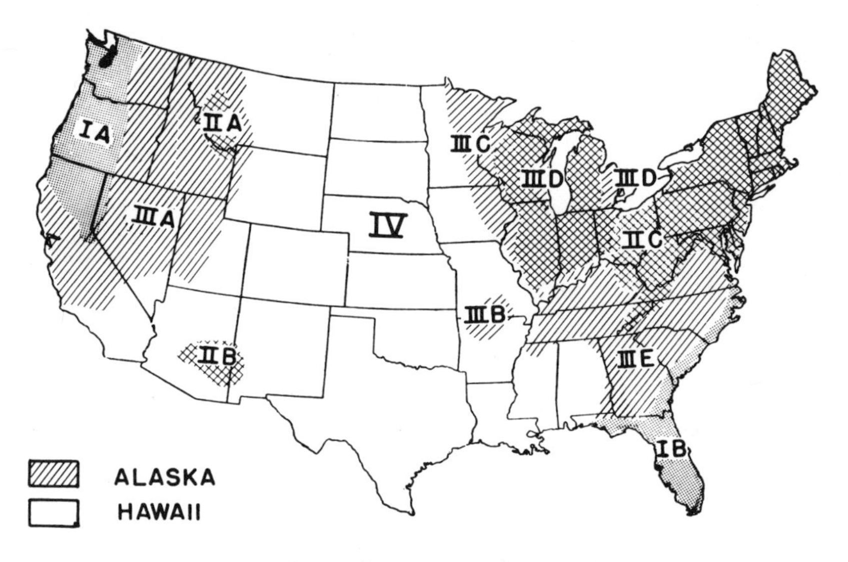

AREA	CROP	NO. OF SAMPLES	MEDIAN CONC.	FREQUENCY DISTRIBUTION (%) OF SAMPLES WITH Se CONCENTRATIONS (ppm.) OF:					
				<0.01 TO 0.05	0.05 TO 0.10	0.10 TO 0.50	0.50 TO 1.0	1.0 TO 5.0	>5
IA	FORAGES	69	0.03	81	15	4	0	0	0
IB	"	26	0.02	89	11	0	0	0	0
IIA	"	14	0.05	50	36	14	0	0	0
IIB	"	11	0.05	36	45	19	0	0	0
IIC	"	187	0.05	65	31	4	0	0	0
IIIA	"	261	0.09	20	31	43	4	2	0
IIIB	"	14	0.05	57	14	22	0	7	0
IIIC	"	39	0.09	20	41	26	13	0	0
IIID	"	27	0.10	26	18	49	7	0	0
IIIE	"	79	0.06	50	23	22	5	0	0
IV	FORAGES	205	0.26	3	10	60	18	9	0
	WHEAT*	856	—	—— 9 ——		22	30	34	5
	FEED GRAIN*	262	—	—— 33 ——			22	38	7

*DATA FROM USDA TECH. BULL. 758. 1941.

USDA 1966

Figure 2. Regional distribution of Se in forages and grains in the United States (ppm dry weight).

they represent arithmetic means of plant samples growing over broad regions that encompass both Zn-deficient and Zn-sufficient soils, are relatively impressive. Within Zn-deficient areas, Zn content of dry plant tissue can drop as low as 6–15 ppm, but in this range plant growth is usually depressed and further decreases are prevented. Soils adversely affected by Zn-rich subsurface waters or mine tailings can produce plants with Zn levels over 100 times this high, as illustrated by the last entries in Table 1.

To give perspective to these values, the effects of certain other variables will be considered. Advancing maturity, for example, can induce a drop in Zn in plant tissue. Loper and Smith (1961) have shown (Table 2) that the Zn content of alfalfa and bromegrass decreases from 36 to 25 and 27 to 18 ppm, respectively, as these crops mature.

The contribution of nutrient interaction to Zn content of crops is compli-

TABLE 1. ZINC CONTENT OF CERTAIN PLANT SPECIES (PPM DRY WEIGHT)

Location	Crop	Comments*	No. of Samples	Zinc Content Mean	References
PENNSYLVANIA	Red Clover	Alluvium	9	22.7	French and others (1957)
	do.	Limestone	53	25.8	
	do.	Shale	167	30.0	
	do.	Sandstone	81	30.5	
	do.	Glacial Till	36	26.6	
	do.	Schist	14	27.2	
	do.	Igneous	5	29.2	
VIRGINIA		Soil Series/*p*H			
	Alfalfa	Penn-Bucks/5.8–7.1	21	28.7	Price and others (1955)
	do.	Tatum-Nasan/6.1–6.5	4	24.6	
	do.	Davidson/5.8–7.1	22	26.5	
VIRGINIA	Alfalfa	Cecil Fine Sandy Loam	4	10.93	Price and Hardison (1963)
	do.	Cecil Clay Loam	2	17.5	
	do.	Appling Fine Sandy Loam	2	13.61	
	Red Clover	Cecil Clay Loam	2	12.0	
VIRGINIA	Orchard Grass	Durham	30	9.8	Price and Hardison (1963)
	do.	Helena	6	10.2	
	do.	Cecil	52	8.8	
	do.	Appling	36	8.6	
	do.	Wilkes	4	9.1	
	do.	Colfax	4	8.8	
	do.	Wickham	6	11.1	

TABLE 1. (CONTINUED)

VIRGINIA	Orchard Grass	Penn-Bucks	2	20.5	Price and others (1955)
	do.	Tatum-Nason	4	23.6	
	Red Clover	Penn-Bucks	3	35.6	
	do.	Tatum-Nason	4	41.8	
	do.	Davidson	10	35.7	
	do.	Cecil	7	35.0	
NEW JERSEY	Red Clover	Peat†	1	107.0	Woltz and others (1953)
	do.	Norton†	1	52.0	
	do.	Squires†	1	42.0	
	do.	Palmyra†	1	32.0	
	do.	Washington†	1	24.0	
	do.	Annandale†	1	18.0	
	do.	Metapeake†	1	8.0	
IDAHO					
Ada Co.	Mixed Forage		10	27.0	Jordan, J. V. (1955)
Bannock Co.	do.		7	20.3	
Butte Co.	do.		2	9.0	
Canyon Co.	do.		8	5.6	
Custer Co.	do.		2	10.0	
Gooding Co.	do.		10	7.7	
Jefferson Co.	do.		2	5.5	
Latch Co.	do.		25	14.1	
Payette Co.	do.		15	22.3	

TABLE 1. (CONTINUED)

Location	Crop	Comments*	No. of Samples	Zinc Content Mean	References
WISCONSIN	Alfalfa	Dane Co.	6	28.0	Loper and Smith (1961)
	Red Clover	(Miami Silt Loam)	6	32.3	
	Bromegrass		6	21.8	
WASHINGTON Yakima Valley	Alfalfa	Normal area	1	16.8	Viets and others (1954)
	do.	Zn-deficient area	1	15.6	
	Red Clover	Normal area	1	13.7	
	do.	Zn-deficient area	1	12.0	
NEW JERSEY	Corn (tassel stage)	Sassafras	1	28.3	Prince (1957)
		Norton	1	57.4	
		Washington	1	37.8	
		Cossayuna	1	12.0	
WASHINGTON Prosser	Corn (silking, 6th node)	Ritzville	1	14.8	Viets and others (1953)
OREGON Hermiston	do.	Ephrata	1	20.1	Viets and others (1953)
WASHINGTON Moses Lake	do.	Ephrata	1	15.7	Viets and others (1953)
SOUTH DAKOTA Huron	Corn (silking, 6th node)	Barnes	1	35.2	Viets and others (1953)
Redfield	do.	Beardon	1	19.2	

TABLE 1. (CONTINUED)

NORTH DAKOTA					
Mandan	do.	Huff	1	36.1	Viets and others (1953)
Bow Bells	do.	Williams	1	24.5	
COLORADO	Corn	Near Burlington	3	11.7	Lindsay (1968) written commun.
	do.	Near Eaton	3	8.8	
PENNSYLVANIA	Corn	Beside tailing ponds	1	270.0	Robinson and others (1947)
	do.	From Zn mines	1	790.0	
	Red Clover		1	1300.0	
	do.		1	640.0	
	do.		1	460.0	
	Ragweed		1	1800.0	
	do.		1	3400.0	
	do.		1	2400.0	
	do.		1	2300.0	
	Horsetail		1	4500.0	
	do.		1	4900.0	
	do.		1	2500.0	
	do.		1	3500.0	
NEW YORK	Spinach	Peat, producing Zn toxicity	1	1110.0	Staker and Cummings (1941)
	Lettuce	do.	1	646.0	

*Names, when given alone, refer to soil series; numbers refer to soil *p*H.
†Greenhouse experiment; all soils limed to *p*H 7.

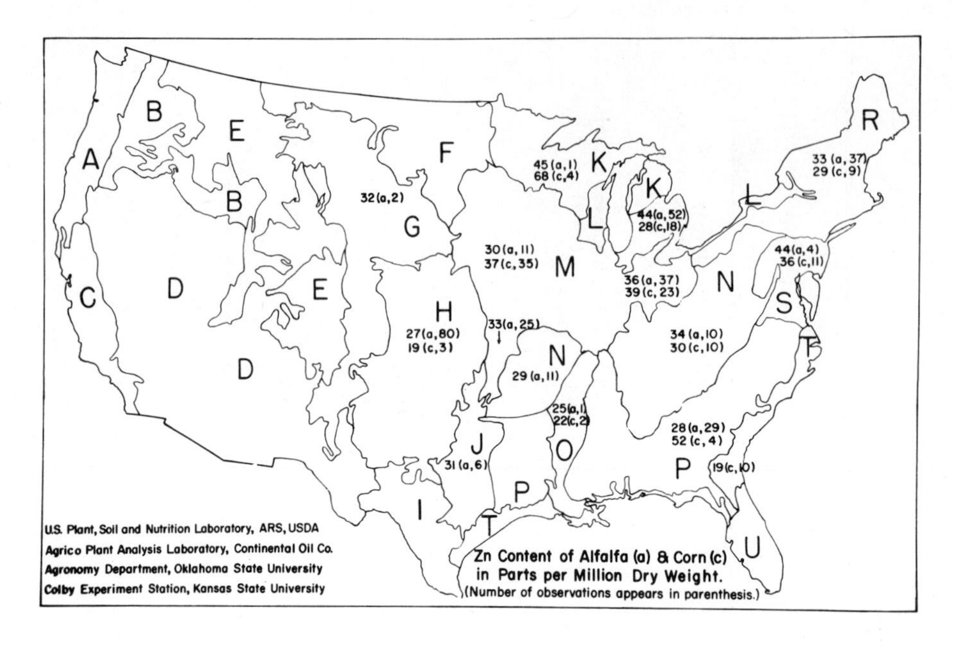

Figure 3. Regional distribution of Zn in corn and alfalfa in the eastern and central United States.

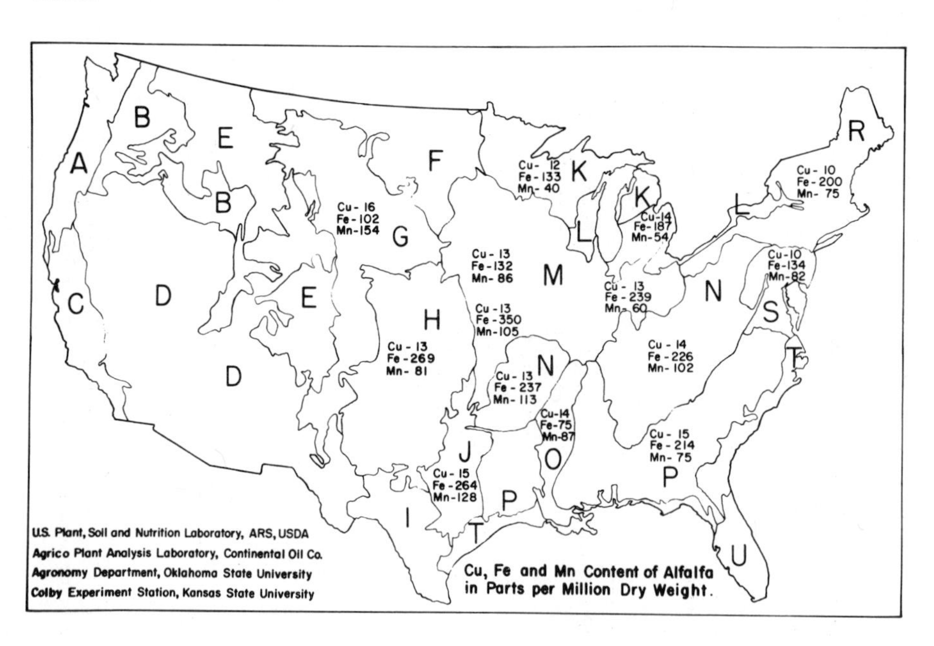

Figure 4. Regional distribution of Cu, Fe, and Mn in alfalfa in the eastern and central United States.

TABLE 2. ZINC CONCENTRATION (PPM DRY WEIGHT) IN ALFALFA AND BROMEGRASS WITH ADVANCING MATURITY (LOPER AND SMITH, 1961)

Alfalfa Stage of Growth	(ppm Zinc)	Bromegrass Stage of Growth	(ppm Zinc)
Succulent	36	Succulent	27
Prebud	31	Root	21
Midbud	29	Head Emerging	24
1/10 Bloom	23	Heading	23
Full Bloom	24	Milk	17
Green Seed Pod	25	Dough	18

TABLE 3. EFFECT OF IRON AND PHOSPHORUS CONCENTRATION ON THE UPTAKE OF ZINC FROM NUTRIENT SOLUTION (WATANABE AND OTHERS, 1965)

Iron in Solution	Phosphorus in Solution	Zinc in Tissue (dry weight)
molar	molar	ppm
4×10^{-5}	1×10^{-4}	57
do.	2×10^{-4}	58
do.	6×10^{-4}	77
8×10^{-5}	1×10^{-4}	46
do.	2×10^{-4}	25
do.	6×10^{-4}	17

cated and large. Iron and phosphorus have particularly profound effects. Watanabe and others (1965) have shown that, in nutrient solution, increasing P in solution can actually increase Zn content of corn when Fe is low, but the same increase of P can sharply decrease Zn in the plant when Fe is high (Table 3). Increasing Fe from 4×10^{-5} M to 8×10^{-5} M when P concentration was high, decreased the Zn content of corn from 77 to 17 ppm in the tissue.

Of particular interest among factors influencing the Zn content of plants are species and varietal differences. Gladstones and Loneragan (1967) determined the Zn content of different plant species grown under the same experimental field conditions. Results are expressed on a dry-weight basis in Table 4. The differences among plant species are striking. Zinc content of different forage species varies by as much as a factor of 4 when grown on the same area.

Although differences among varieties are not as large as among species, they are also impressive. Table 5 gives the Zn content of inbred lines of corn grown in Kentucky and a group of corn hybrids grown in Oklahoma. The Zn content varies approximately three-fold in the first case and more than two-fold in the second. Zinc content was not a factor in developing these hybrids, but differences in species and varietal content of elements underline the potential of plant breeding for changing the composition of foods and feeds.

Differences in plant composition also direct attention to food selection as a means of altering the Zn supply to animals and man. Schroeder and others (1967) tabulated the Zn content of a variety of foods produced under different conditions. Results are given on a fresh-weight basis in Table 6. Meat contains over 10 times as much Zn as most fresh vegetables.

What conclusions can be drawn from this information concerning problems of human health? Can data of this kind serve as a guide in allocating our limited resources to such problems? Before examining the significance of particular data, such as presented here, consider first, in a more general way, what

TABLE 4. ZINC CONTENT OF DIFFERENT PLANT SPECIES GROWN UNDER THE SAME EXPERIMENTAL FIELD CONDITIONS. RESULTS ARE EXPRESSED AS PPM ON A DRY-WEIGHT BASIS (GLADSTONES AND LONERAGAN, 1967)

Species and Variety	ppm
French serradella (*Ornithopus sativus*)	51
W. A. serradella (*Ornithopus compressus*)	42
W. A. blue lupine (*Lupinus digitatus*)	38
Yellow lupine (*Lupinus luteus*)	44
Rose clover (*Trifolium hirtum*)	28
Field peas (*Pisum arvense*)	22
New Zealand blue lupine (*Lupinus angustifolius*)	27
Subterranean clover (*Trifolium subterraneum*) var. Bacchus March	21
White lupine (*Lupinus albus*)	37
Subterranean clover (*Trifolium subterraneum*) var. Clare	18
Subterranean clover (*Trifolium subterraneum*) var. Yarloop	22
Purple vetch (*Vicia atropurpurea*)	18
Barrel medic (*Medicago truncatula*)	19
Cape-weed (*Cryptostemma calendula*)	39
Erodium (*Erodium botrys*)	21
Cereal rye (*Sercale cereale*)	28
Barley (*Hordeum sativum*)	18
Wheat (*Triticum aestivum*)	14
Oats (*Avena sativa*) var. Ballidu	13
Oats (*Avena sativa*) var. Avon	12
Wimmera ryegrass (*Lolium rigidum*)	22
Silver grass (*Vulpia* sp.)	20
Ripgut brome (*Bromus rigidus*)	19
Soft brome (*Bromus mollis*)	12

might be accomplished by knowing more about the geographic distribution of crops in relation to trace-element content.

(1) It would permit a comparison of the incidence of endemic diseases of unknown origin with a possible cause. Although this approach does little to prove a cause and effect relationship, it can serve as an important guide for further work.

(2) If a cause and effect relationship is known, knowledge of the geographic distribution of an element may be the best way to delineate the extent of the area that is affected by the disease.

(3) Through such information, it may be possible to define locations of greatest interest for more intensive field study.

(4) Understanding the agronomic complications involved in the composition of crops in different areas should help in determining the most practical way to control the problem. Alteration of the supply of an element to the human population might be approached by way of agronomic activities such as additions to or alterations in the soil (which will subsequently affect plant composition), or by plant breeding to select varieties that accumulate or exclude elements of concern; or, through nonagricultural practices such as food enrichment as is done with I, Fe, and Ca, or water supplementation as

TABLE 5. ZINC CONTENT OF INBRED CORN LINES GROWN IN KENTUCKY, AND CORN HYBRIDS GROWN IN OKLAHOMA IN PPM DRY WEIGHT (MASSEY AND LOEFFEL, 1966; AND LOCKMAN)

Inbred Line	Zinc in Kernels	Corn Hybrid	Zinc in Leaves
	ppm		ppm
Ky 211	38.4	Penn 507	63
H 21	26.3	U.H. 108 (Can)	56
Oh 45	25.9	DeKalb XL45	55
Ky 122	25.3	Pioneer 3304	52
W F 9	25.0	D.P.H. X77	50
Ky 209	24.9	Mark 227	50
38 - 11	24.8	Mark 397	49
M 14	24.8	PAG SX29	47
B 2	23.9	Funk G 4641	45
33 - 16	23.7	Mark 218	43
Oh 41	23.6	Pioneer 3306	42
CI 21 E	23.6	D.P.H. 225	41
Ky 214	23.3	Funk G 4601	39
Ky 215	22.9	Funk G 4401	26
L 317	22.6		
K 155	22.5		
Ky 213	22.1		
Ky 27	21.8		
T 8	21.1		
Ky 201	20.8		
K 56	19.7		
Hy	19.2		
Ill L	19.1		
Oh 7 B	18.7		
K 64	17.5		
Oh 07	17.1		
Ky 36 - 11	16.4		
CI 64	16.2		
Oh 29	15.5		

TABLE 6. ZINC CONTENT OF SOME COMMON FOODS EXPRESSED ON A FRESH-WEIGHT BASIS (SCHROEDER AND OTHERS, 1967). VALUES REPRESENT AVERAGES OF MANY INDIVIDUAL FOODS

Food	Zinc ppm
Seafood	17.5
Meat	30.6
Milk	0.14–0.45
Cereals and Grain	19.7
Beans and Green Vegetables	6.2
Other Vegetables	2.3
Oils and Fats	8.4
Nuts	34.2

carried out by some localities with F, or through medication in the form of injections or pills.

(5) Through a better understanding of natural distribution of elements in plants and their removal from soils, it would be possible to anticipate more accurately the consequences of specific forms of pollution.

How does available information on Zn contribute then, to this kind of scheme?

TABLE 7. ZINC CONCENTRATIONS IN WHOLE BLOOD FROM 19 CITIES IN THE UNITED STATES

Source	No. Samples	Zinc µg/100 ml	
		(mean)	(range)
Canandaigua, N.Y.	20	1,008	579–1,987
Montpelier, Vt.	10	920	738–1,183
Missoula, Mont.	29	789	356–1,173
Geneva, N.Y.	15	766	159–1,285
Red Bluff, Calif.	12	726	517–1,113
Billings, Mont.	12	578	190–1,267
Muncie, Ind.	10	560	444– 711
Lima, Ohio	10	503	341– 674
Lubbock, Tex.	12	496	301– 845
Spokane, Wash.	10	493	321– 667
Jacksonville, Fla.	9	430	180– 591
El Paso, Tex.	12	421	66–1,327
Little Rock, Ark.	10	368	284– 464
Rapid City, S.D.	22	367	240– 580
Lafayette, La.	10	336	88– 448
Meridian, Miss.	10	325	199– 468
Fargo, N.D.	10	319	222– 438
Phoenix, Ariz.	10	264	80– 385
Cheyenne, Wyo.	10	198	60– 420
Summary	243	530	60–1,987

Certainly despite, if not because of, the many complications affecting Zn uptake, regional variations appear to exist in the Zn content of crops. Furthermore, variations within broad regions must certainly accentuate the differences reported here. Some indication of this is given in Table 1 where individual areas in Idaho have means below 6 ppm Zn in mixed forage.

Geographic variations in Zn content of plants are not so large, however, that food selection and movement from one area to another might not conceivably mask local variations in the Zn content of plant tissue. In view of this, it is of interest to examine the variations in Zn content of human blood. From Table 7 (taken *from* Kubota and others, 1968), it is apparent that all of the locations having a mean Zn concentration in the blood of below 500 µg/100 ml are from areas associated either with low levels of Zn in alfalfa (and, in most cases, corn), or areas of Zn deficiency in plants, or both.

From these considerations, it is reasonable to expect that endemic patterns of Zn-responsive disorders could be observed in human beings, with appropriate selection of locations from known areas of Zn deficiency.

With regard to increasing the supply of Zn to the human population, the many nutrient interactions, low assimilation by monogastric animals, and lack of incentive by the farmer to supply Zn beyond the needs of his own crops or farm animals makes the agronomic approach unattractive. There is sufficient variation in Zn content among plant varieties to make breeding programs worthy of study. On the other hand, food enrichment has proven to be an effective means of supplementing the human diet with I, Fe, and Ca. The expense is minimal, but it does require abundant assurances that the supplements are, in fact, beneficial and that no harmful effects result from the concentrations that would be added. Supplying Zn by medication has the advantage of having more personal choice, but it would not be as satisfactory in meeting the needs of an entire population.

Problems related to those elements that are found in excess pose an added challenge. Whereas simple deficiencies can be controlled by the addition of the element in question at any of several points in the food chain, in the case of toxicities, no such obvious solutions are evident; but agronomic manipulation offers considerable opportunity for reducing the supply of such elements to man by altering their availability in the soil or changing plant response through breeding programs.

In the final analysis, we must look to the time when greater attention is paid to levels of all elements important to human nutrition. In this paper we have examined some of the factors that may contribute to this change, and the agricultural implications that may result from it.

REFERENCES CITED

Beeson, K. C., 1959, Plant and soil analysis in the evaluation of micronutrient element status, *in* Mineral Nutrition of Trees: Duke Univ. Forestry School Bull. 15, p. 71–80.

French, C. E., Smith, C. B., Fortman, H. R., Pennington, R. P., Taylor, G. A., Hinish, W. W., and Swift, R. W., 1957, Survey of ten nutrient elements in Pennsylvania forage crops: Pennsylvania Agr. Expt. Sta. Bull. 624, 28 p.

Gladstones, J. S., and Loneragan, J. F., 1967, Mineral elements in temperate crop and pasture plants. I. Zinc: Australian Jour. Agr. Research, v. 18, p. 427–446.

Jordan, J. V., 1955, Protein and mineral content of forage legumes and grasses in Idaho: Idaho Agr. Expt. Sta. Bull. 245, p. 1–8.

Kubota, J., 1968, Distribution of cobalt deficiency in grazing animals in relation to soils and forage plants of the United States: Soil Sci., v. 106, p. 122–130.

Kubota, J., Allaway, W. H., Carter, D. L., Cary, E. E., and Lazar, V. A., 1967, Selenium in crops in the United States in relation to selenium-responsive diseases in animals: Agr. Food Chemistry, v. 15, p. 448–453.

Kubota, J., Lazar, V. A., and Losee, F., 1968, Copper, zinc, cadmium, and lead in human blood from 19 locations in the United States: Arch. Environ. Health, v. 16, p. 788–793.

Loper, G. M., and Smith, D., 1961, Changes in micronutrient composition of the herbage of alfalfa, medium red clover, ladino clover, and bromegrass with advance maturity: Wisconsin Agr. Expt. Sta. Research Rept. no. 8, 19 p.

Massey, H. F., and Loeffel, F. A., 1966, Variation in zinc content of grain from inbred lines of corn: Agronomy Jour., v. 58, p. 143–144.

McClendon, J. F., 1939, Iodine and the Incidence of Goiter: Minneapolis, The University of Minnesota Press, 126 p.

Price, N. O., and Hardison, W. A., 1963, Minor element content of forage plants from the central Piedmont region of Virginia: Virginia Agr. Expt. Sta. Bull., no. 165, 15 p.

Price, N. O., Linkows, W. N., and Engel, R. W., 1955, Minor element content of forage plants and soils: Agr. Food Chemistry, v. 3, p. 226–229.

Prince, A. L., 1957, Influence of soil type on the mineral composition of corn tissues as determined spectrographically: Soil Sci., v. 83, p. 399–405.

Robinson, W. O., Lakin, H. W., and Reichen, Laura, 1947, The zinc content of plants on the Friedensville zinc slime ponds in relation to biogeochemical prospecting: Econ. Geology, v. 42, p. 572–582.

Schroeder, H. A., Nason, A. P., Tipton, Isabel H., and Balassa, J. J., 1967, Essential trace metals in man: Zinc. Relation to environmental cadmium: Jour. Chronic Disease, v. 20, p. 179–210.

Staker, E. V., and Cummings, R. W., 1941, The influence of zinc on the productivity of certain New York peat soils: Am. Soil Sci. Soc. Proc., v. 6, p. 207–214.

Viets, F. G., Jr., Boawn, L. C., and Crawford, C. L., 1954, Zinc contents and deficiency symptoms of 26 crops grown on a zinc-deficient soil: Soil Sci., v. 78, p. 305–316.

Viets, F. G., Jr., Boawn, L. C., Crawford, C. L., and Nelson, C. E., 1953, Zinc deficiency in corn in central Washington: Agronomy Jour., v. 45, p. 559–565.

Watanabe, F. S., Lindsay, W. L., and Olsen, S. R., 1965, Nutrient balance involving phosphorus, iron and zinc: Am. Soil Sci. Soc. Proc., v. 29, p. 562–565.

Woltz, S., Toth, S. J., and Bear, F. E., 1953, Zinc status of New Jersey soils: Soil Sci., v. 76, p. 115–122.

MANUSCRIPT RECEIVED BY THE SOCIETY OCTOBER 8, 1969

Printed in U.S.A.

THE GEOLOGICAL SOCIETY OF AMERICA, INC.
MEMOIR 123, 1971

Zinc Deficiency in Delayed Healing and Chronic Disease

WALTER J. PORIES
AND
WILLIAM H. STRAIN
Cleveland Metropolitan General Hospital
Case Western Reserve University School of Medicine
Cleveland, Ohio
AND
CHARLES G. ROB, M.D.
School of Medicine and Dentistry
University of Rochester
Rochester, New York

ABSTRACT

Recognition of zinc deficiency as a problem in man has taken a long time. Although the metal was shown to be essential for the development and the growth of *Aspergillus niger* in 1869, of plants in 1914, and of animals in 1934, an essential role of zinc in man was considered to be unlikely until recently. Within the past few years, however, investigators have demonstrated, in rapid succession, that zinc deficiency is common in man, that this deficiency is a critical factor in some forms of impaired growth, delayed healing, and chronic disease. The development of systematic zinc therapy for man offers a new approach to the treatment of growth problems, indolent wounds, and other conditions associated with zinc deficiency.

CONTENTS

ZINC DEFICIENCY IN CROPS

Zinc deficiency was first recognized by workers in agriculture (Stiles, 1961). Zinc applications were found to be beneficial on field and vegetable crops in Florida as early as 1927 and on citrus crops in California in 1932. Today, the deficiency of zinc in soils is recognized to be more common, world-wide, than that of any other trace element. In the United States (Fig. 1), zinc deficiency is known in 32 different states (Berger, 1962). Fertilization with only a few pounds of zinc per acre has produced remarkable increases in productivity.

Zinc deficiency results in a variety of diseases in crops. Examples are: "white bud" of corn, "little leaf" of apple and pear, "rosette" of pecan, and

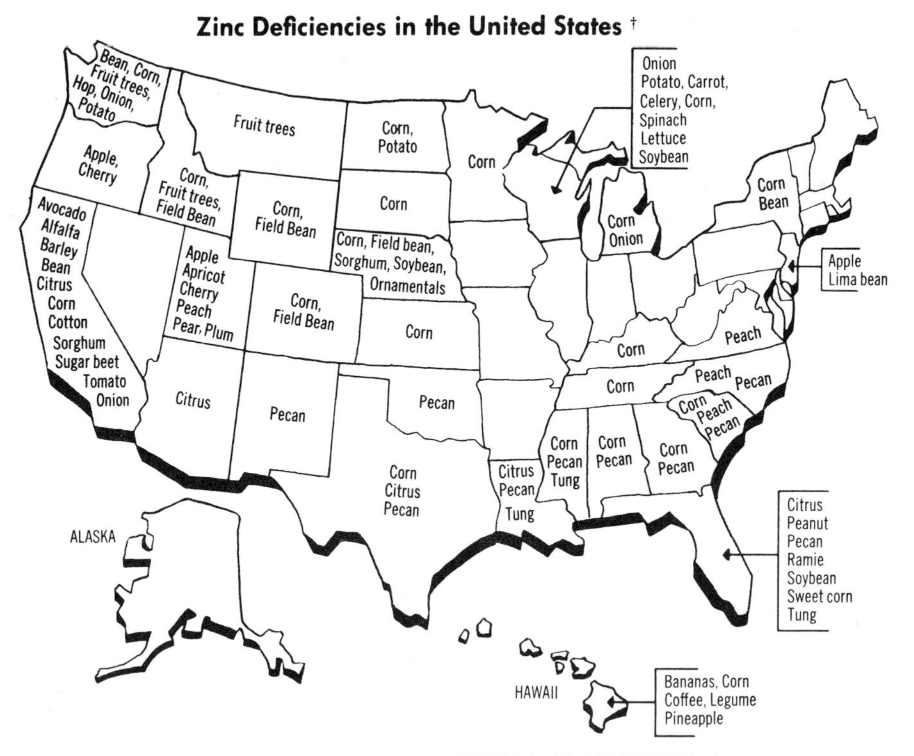

†Reprinted (with minor additions*) from the JOURNAL OF AGRICULTURAL AND FOOD CHEMISTRY, Vol. 10, No. 3, May/June 1962.

*Additions (field bean in Idaho, Wyoming, Colorado; corn in Wyoming, Colorado)

Figure 1. Zinc deficiencies have been reported in 32 of the United States shown above. The field crops most frequently stunted and reduced in yield include: corn, cotton, beans, onions, potatoes, and tomatoes. Citrus and other fruits, and nuts are dependent on an adequate supply of zinc for maximum yield. The quality of the plant protein is also improved.

"mottle leaf" of citrus. The results are low yields, poor seed formation, and, frequently, total crop failure.

Much of our understanding of the biochemical functions of zinc comes from studies in plants (Price, 1963; Stiles, 1961). Protein synthesis, and, more specifically, tryptophan synthesis (Tsui, 1948), is depressed in zinc deficient *Mycobacterium smegmatis,* in which the primary effect of the deficiency appears to be interference with the formation of RNA and, consequently, of DNA (Schneider and Price, 1962; Wacker, 1962). The impairment of carbohydrate metabolism in fungi and higher plants deficient in zinc appears to result from failure of interaction between the element and thiamine (Tomita, 1961). The remarkable effects of zinc on growth and maturation continue to be unexplained.

Zinc deficiency in plants develops from three main soil conditions: (1) low content of total zinc; (2) unavailability of bound zinc; and (3) poor management practices (Stiles, 1961; Viets, 1963). Total zinc is low in highly leached soils, such as those found in many coastal areas. The element is made unavailable by alkaline soils, by a high content of organic matter, and by a high concentration of magnesium or phosphate, often found in clay soils. More recently, heavy fertilization of soils with phosphates and nitrogens has contributed greatly to the zinc deficiency of soils and, thus, of crops.

It is not surprising that such widespread zinc deficiency in soils should lead to deficiencies in our crops and our food animals. Yet it is surprising that it took so long to discover the problem and to apply the obvious solution of adding the missing zinc to soils and feeds. It is obvious that if our vegetable and animal food sources are deficient, man also may be subject to a deficiency in this element.

ZINC DEFICIENCY IN ANIMALS

Zinc is now recognized to be an essential element for all animals (Underwood, 1962). It is required in only minute concentrations, such as 20 to 100 ppm, yet even slight or moderate deficiencies can retard growth, lower feed efficiency, and inhibit general well-being. Characteristic signs of zinc deficiency in animals include disorders of the bones, the joints, and of the skin; delayed healing; and loss of fertility. Zinc deficiency, in severe forms, has been a cause of death in American animal herds. The availability of zinc from animal rations is also calcium dependent.

Swine

Tucker and Salmon (1955) at Alabama Polytechnic reported on the beneficial effects of supplemental dietary zinc for swine. They found that the addition of zinc to peanut flour converted this toxic food into an acceptable ration and prevented the severe parakeratosis and deaths which resulted from the peanut-flour diet. At that time, parakeratosis was known to occur in about 20 percent of growing pigs in confinement feeding. Balancing calcium with zinc was a new concept from this investigation, and further work indicates that this balance is important in all animals.

The reports from Hoekstra and others (1956) at Wisconsin are particularly informative. They studied the effects of increasing the calcium content of the diet while holding the zinc level constant and, conversely, of increasing the zinc level while maintaining a constant calcium level. Data on some of these calcium:zinc relationships are given schematically in Figure 2A. Increasing the calcium content of the diet caused all the pigs to have parakeratosis, but increasing the zinc content prevented the dermatitis.

The effect of zinc in promoting weight gain in swine was brought out graphically in the recent work carried out by Bell and associates (Berry and others, 1966) at the University of Tennessee. As shown in Figure 2B, a pig raised on a low-zinc–high-calcium diet gained only 68 grams per day and had many skin lesions. After 16 weeks on this poor diet, the stunted animal was placed on high-zinc–low-calcium rations and promptly gained 1022 gms per day, and

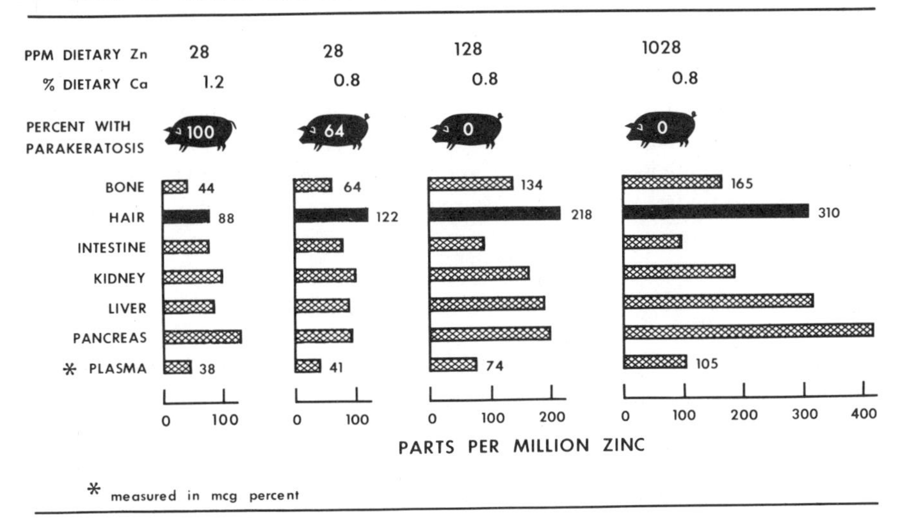

Figure 2A. The effect in swine of varying the intake of calcium and zinc on the occurrence of parakeratosis and on the zinc levels of various tissues and hair. Stratified groups of six pigs selected from 30 weanlings were placed for 20 weeks on diets varying only in the calcium and zinc content as shown in the heading of the chart. The incidence of parakeratosis decreased from 100 to 0 percent as the calcium content of the diet was reduced from 1.2 to 0.8 percent, and the zinc content was increased from 28 to 128 ppm. With the change in the calcium:zinc ratio of the diet, there were marked changes in the zinc level of the tissues and hair. The hair zinc levels reflect these changes better than the serum zinc levels (adapted *from* W. G. Hoekstra, 1956, with permission).

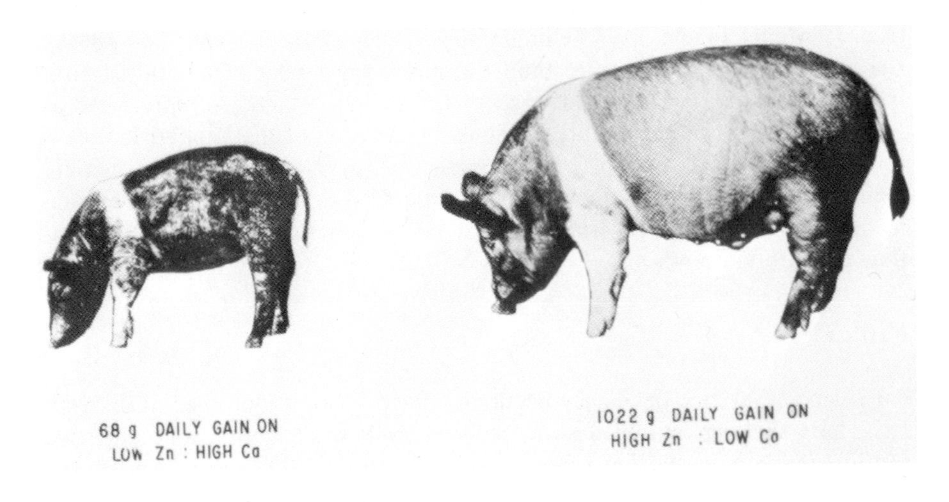

Figure 2B. A pig stunted by eating a low-zinc–high-calcium diet for 16 weeks becomes normal on a high-zinc–low-calcium diet after 9 weeks. The recovery was very rapid when 71 ppm of zinc was added to the initial diet (adapted *from* M. C. Bell [Berry and others, 1966], with permission).

became normal in size for his age after nine weeks; and, at the same time, his skin lesions healed.

The demonstration that zinc deficiency is a common problem in swine was a major turning point in zinc research. Although zinc deficiency had been produced in rats in 1934 (Todd and others, 1934), the studies in swine were the first evidence that zinc deficiency occurred naturally in domestic animals and that these deficits were important to our economy. Today, the routine addition of 50 to 100 ppm of zinc to swine food has increased feed efficiency as much as 25 percent.

Poultry

O'Dell and Savage (1957) at the University of Missouri published data on the beneficial effects of zinc supplementation on the growth of poultry. Publications from nearly every other agricultural college followed. About six million tons of commercial poultry rations are now supplemented with 35 to 70 ppm of zinc, again indicating the widespread lack of zinc in this country. The added zinc increases feed efficiency as much as 25 percent and prevents the development of large hocks, dermatoses, and frizzly feathers.

We studied the calcium-zinc interrelationships in chickens in a cooperative study with Zeigler and Scott (Zeigler and others, 1962) of Cornell University in 1961. Pullets were housed in lacquered cages and fed a basal soybean diet. Although the diet was analytically adequate in zinc, little of the zinc was available to each chick, because of the chelation of the metal by natural phytates. Accordingly, the diet was supplemented with varying levels of zinc, 5, 10, 15, or 60 ppm, as shown in Figure 3A. Representative pullets were photographed at the end of 10 weeks when they showed the differences in weight recorded under Figure 3A. The pullet on the highest supplement of 60 ppm of zinc weighed four times more than the pullet on 5 ppm of additional zinc. Roentgenograms of a leg from each of the pullets, which are reproduced as Figure 3B, showed that maturation and calcification of the joint structures is proportional to the amount of zinc added to the diet. The studies demonstrated that adequate amounts of zinc are required to produce proper development and calcification of joints and that zinc is essential for growth, maturation, and protein synthesis.

Cattle

Evidence that zinc deficiency occurs in cattle was not published until 1960. Itch, hair slicking, wrinkled skin, reduced milk productivity, and impaired fertility in dairy cattle have been attributed to zinc deficiency. Miller and Miller (1960) at the University of Georgia induced zinc deficiency by synthetic diets low in zinc and by the addition of cadmium to the diet. If soybean-corn rations are supplemented with 18 to 42 ppm of zinc, feed efficiency is increased as much as 15 percent.

Figure 3A. The effect on growth of varying zinc levels is shown in these 10-week-old pullets (White Cornish × White Rock). The zinc content of the rations, the weights after 10 weeks, the zinc content of various portions of the feathers, and the zinc content of femur ash were, from left to right:

Zinc, ppm	5	10	15	60
Weight, grams	355	680	840	1560
Feather zinc, ppm				
Tip	8.7	11.4	11.1	15.7
Middle	22.4	24.0	26.1	39.2
Shaft End	29.9	41.3	33.0	27.4
Femur ash zinc, ppm	140	226	180	360

Pullets on low zinc rations show poor feathering, enlarged hock joints, extreme weakness, and difficulty in standing erect.

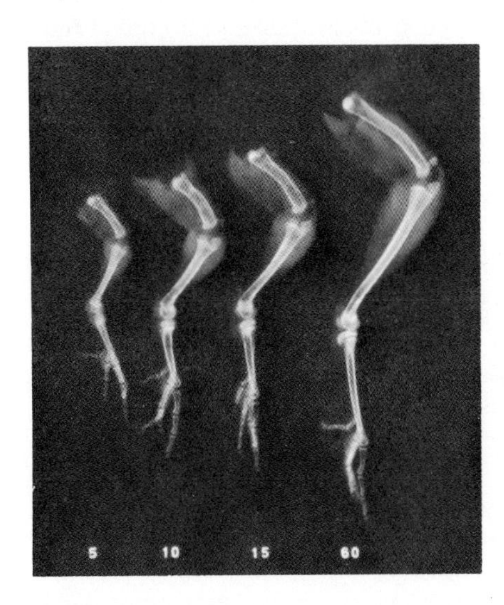

Figure 3B. Radiographs of the legs from the pullets at 10 weeks showing the effects of zinc deficiency on calcium deposition, and on shortening and thickening of the long bones. Comparing lengths of femora with 60 ppm as the normal, the 5 ppm is slightly less than half, the 10 ppm slightly more than half, and the 15 ppm approximately two-thirds that of the normal. The difference in caliber of the long bones is not proportionately similar to the difference in lengths; only the 5 ppm is significantly smaller in diameter. The striking difference is centered in the area of the epiphysis, where marked retardation in development and ossification is present in the most deficient bird and progressively less marked as the level of supplementation is increased. Cortical bone is thinner in the deficient birds, and there is less evidence of endosteal reconstruction in the cancellous portion.

Miller and others (1965) also confirmed that adequate zinc stores are needed for normal wound repair. He demonstrated that zinc-deficient cattle heal poorly and develop a halo of parakeratosis about the healing site. Although the addition of zinc to the feed produced rapid normal healing, the addition of excess zinc to a zinc-sufficient diet produced no acceleration of healing rates.

We recently investigated the zinc content of milk. Archibald (1958) at the University of Massachusetts and Miller and others (1966) at the University of Georgia showed that the zinc content of milk varies up to 8 ppm with increasing dietary intake of zinc. We analyzed dried milk from various areas of New York State and have found that the content of zinc is only 1.5 ppm and not the 4 ppm or higher that follows from a good intake of zinc. Similar low zinc levels have been reported in compilations by the American Dry Milk Institute (Anonymous, 1964) of the dried milk collected throughout the United States. The increasing practice of supplementing dairy feeds with adequate zinc will produce milk with a higher zinc content, and this may be a significant health measure.

ZINC DEFICIENCY IN MAN

Because of the prevalency of zinc deficiency in soils, crops, and food animals, it was only a matter of time before this deficiency would be identified in man. The first report was published by Vallee and others (1956), who found a mean serum zinc level of 66.5 ± 19 mcg percent in post-alcoholic cirrhosis patients compared to a mean of 121 ± 19 mcg percent for normal patients. In 1962, we (Pories and others, 1966) found that male patients with atherosclerosis were also severely zinc deficient with hair zinc levels of 62.2 ppm compared to normal range of 125 to 150 ppm. Our knowledge of the zinc-deficiency syndrome in man was greatly extended by Prasad and others (1963), who studied extensively a series of Iranian and Egyptian dwarfs with profound zinc deficits.[1] These patients had hepatosplenomegaly, hypogonadism, and marked growth retardation. Their plasma zinc level averaged 73 ± 6 mcg percent and their hair zinc level 54.1 ppm, within the range of the patients with cirrhosis and atherosclerosis (*see* above). When the dwarfs were treated with oral zinc sulfate and placed on good diets, the hepatosplenomegaly receded, the genitalia developed, and other normal changes of puberty took place (Figs. 4A, B, and C). The zinc levels of their plasma and hair rose to normal. The shortest dwarf, who was 20 years old and 39 inches tall, grew five inches in 14 months on zinc sulfate therapy. Today, it is evident that zinc deficiency is common in man and especially so in hospital patients with chronic disease (Sullivan and others, 1968).

ZINC AND DEVELOPMENT

The organism is most sensitive to zinc deficits during early stages of development and growth. When zinc deficiency is produced in hen's eggs and in

[1]Subsequent work has clouded these results somewhat and the observations have become controversial. Further studies made with Egyptian male dwarfs (Coble and others, 1966; Carter and others, 1969) showed no effect of supplementation with zinc, as compared with placebos, on height, weight, or sexual development. Other investigations on malnourished boys (Caggiano and others, 1969; Ronaghy and others, 1969) provide additional evidence for the hypothesis that zinc deficiency is an important etiological factor in growth retardation and delayed sexual maturation in males.

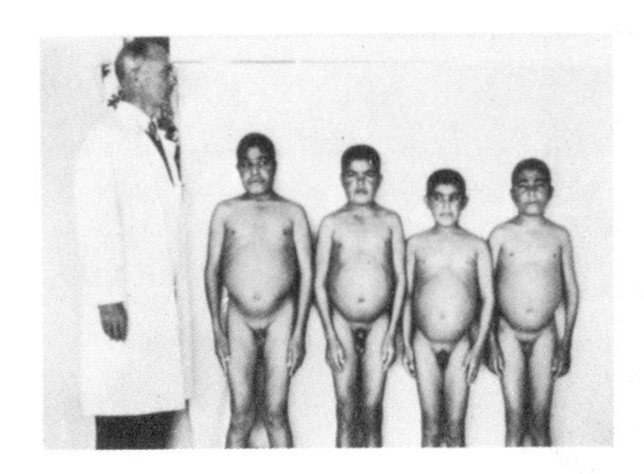

Figure 4A. Clay-eaters from Iran (geophagia). From left to right: age 21, height 4 ft, 11-1/2 in.; age 18, height 4 ft, 9 in.; age 18, height 4 ft, 6 in.; age 21, height 4 ft, 7 in. All patients presented the following clinical features: severe iron deficiency anemia, hepatosplenomegaly, short stature, and marked hypogonadism. Subsequent work has indicated that the failure to grow was due to zinc deficiency. The physician standing at the left of the dwarfs is 6 ft tall.

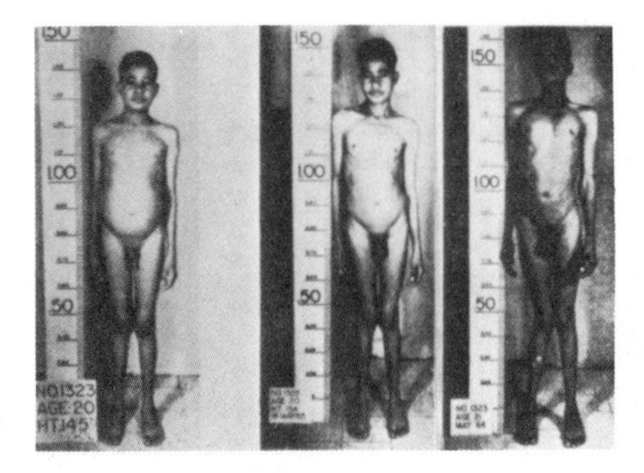

Figure 4B. Growth of Egyptian dwarf produced by zinc supplement in 18 months. His height increased nearly 8 in., pubic hair grew, and his genitalia became adult in size (*from* A. S. Prasad, 1963, with permission).

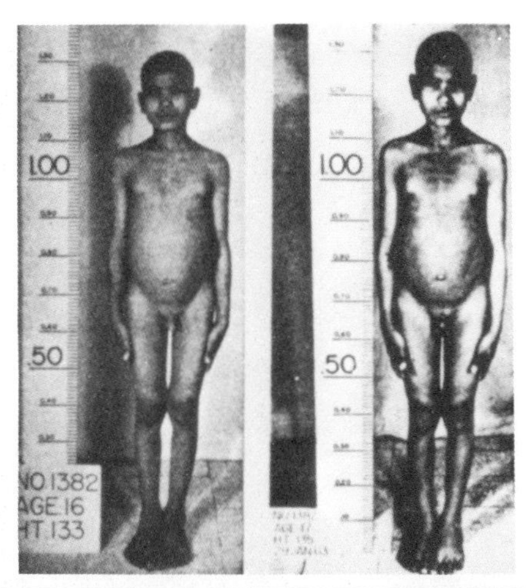

Figure 4C. Egyptian dwarf who received iron in place of zinc supplement. Pubic hair did not grow and the external genitalia did not develop. Scale is in meters (*from* A. S. Prasad, 1963, with permission).

pregnant rats, striking skeletal and other birth defects develop in the offspring. The grossly deformed chick embryos with beak, eye, limb, and trunk defects shown in Figures 5A, B, and C were obtained by Supplee's group (Blamberg and others, 1960) at Maryland and Hoekstra's group (Kienholz and others, 1961) at Wisconsin. The current work of Hurley and Swenerton (1966) and Swenerton and Hurley (1968), at the University of California at Davis, has established that pregnant rats placed on zinc-deficient diets produce litters with brain, eye, kidney, and skeletal defects. Their results with the rat, a mammal, are suggestive of the birth defects produced in man by the drug, thalidomide.

The sensitivity of the developing animal to zinc deficiency led us to examine the zinc levels of growing infants during the first year of life. We (Strain and

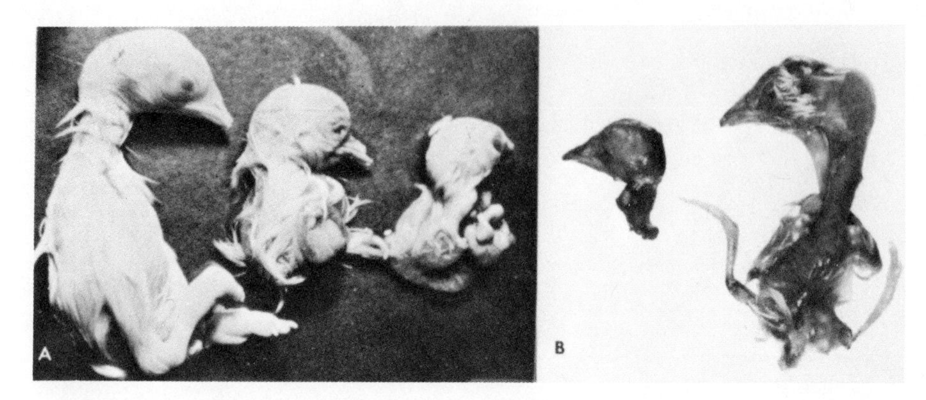

Figure 5. Birth defects in embryos from egg laid by hens fed zinc-deficient diets. A. Incubated 18 days; left, normal embryo; center and right, skeletal, beak and eye defects (*from* W. C. Supplee *in* Blamberg and others, 1960). B. Incubation 21 days and alive at normal hatching time; left, complete absence of skeleton below the cervical vertebra; right, severe skeletal changes and edema of the pipping muscle (*from* W. G. Hoekstra, 1956; both photographs published with permission).

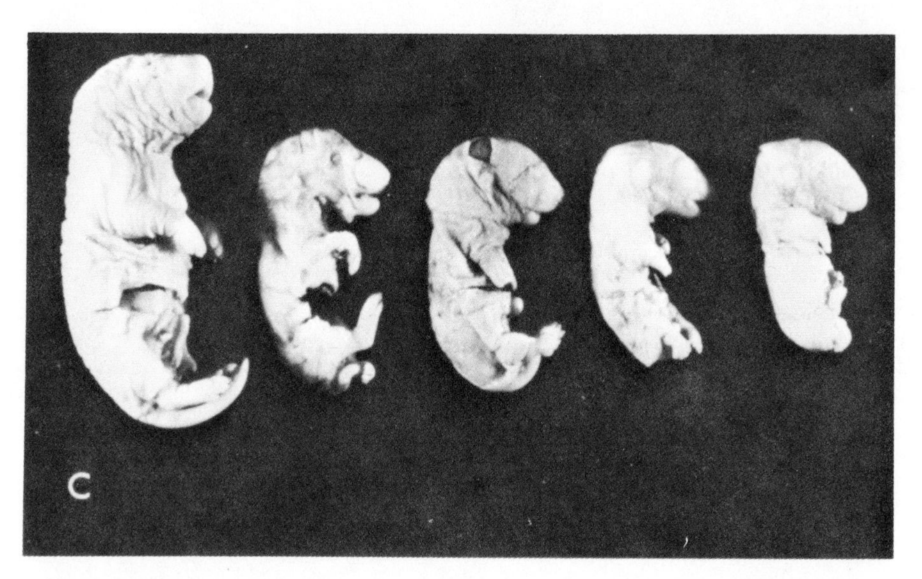

Figure 5C. Rat fetus on left from normal mother compared to fetuses from zinc-deficient mothers. Zinc-deficient mothers produce smaller offspring with abnormally shaped heads, clubbed feet, fused or missing digits, short lower jaw, short or absent tail, brain defects, and kidney abnormalities (*from* Hurley and Swennerton, 1966, with permission). Some sections have been removed.

others, 1966) found in 8 of 12 babies that the zinc content of hair fell sharply from a mean birth value of 138 ± 16 ppm to levels of 82 ppm or less. Recent studies by Fomon (1968, written commun.), at the University of Iowa have confirmed this work and have also shown negative zinc balances in young infants. It seems reasonable to assume that marginal zinc stores in the growing child might be a limiting factor in its normal development and growth.

ZINC AND HEALING

The empiric use of zinc to promote healing dates back to the early Egyptians who applied it topically in the form of calamine. Since then, zinc oxide, zinc sulfate, and zinc stearates in the form of powders, salves, and ointments have continued to be used widely and effectively in the topical treatment of skin lesions.

The Beneficial Contaminant

Our interest in zinc began in the early nineteen fifties with a laboratory accident. We (Strain and others, 1953) were studying wound-healing in rats and trying to control the rate of repair by the addition of various amino-acid analogues to the diet. Among these was beta-phenyllactic acid, which had been added to the feed with the expectation that it would delay repair. To our surprise, it definitely accelerated healing. We repeated the work several times, and in every instance healing was promoted. Because the beta-phenyllactic acid had been obtained from a commercial source and seemed impure, we prepared pure beta-phenyllactic acid ourselves. But the pure compound did not alter the healing rate, and as zinc was used in the chemical preparation of the commercial product, we suspected that zinc was responsible for the accelerated healing.

A subsequent series of ten experiments proved that zinc was the beneficial contaminant. Young and old rats of both sexes were wounded by excising a full thickness, 3 cm circle of dorsal skin, and the closure of the wounds was followed photographically (Fig. 6). On standard commercial diets, the young female rats healed within 30 days, the young male and old female rats within 45 days, but the old male rats required as much as 115 days. The addition of zinc salts to the diets accelerated the healing of the old males as much as 40 percent and the young males and old females as much as 20 percent.

The study showed that the addi-

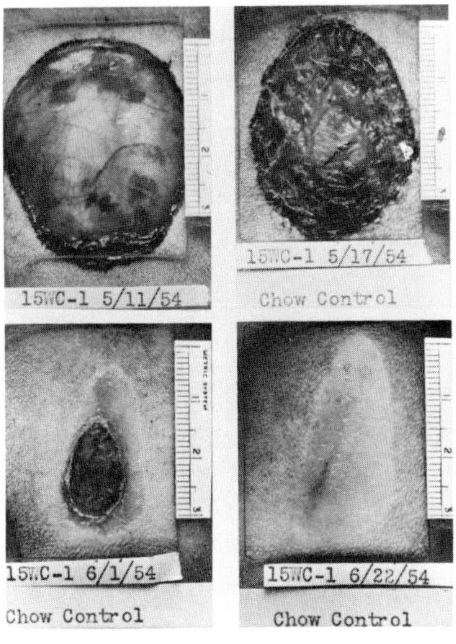

15WC-1 5/11/54

15WC-1 5/17/54
Chow Control

15WC-1 6/1/54
Chow Control

15WC-1 6/22/54
Chow Control

Figure 6. Characteristic phases of healing shown by a young male rat are: (A) Wound formed by excising skin dilates from the original area from 7.07cm^2 to 10.5 cm^2 and (B) then contracts to the original size during the next six days. As healing continues, scar tissue forms (C) until the wound is healed (D). Healing is slower in male than in female rats, and the area of scar tissue is larger and more oval in the male. The healing time is far longer in old males than in the other rats.

tion of easily available zinc could accelerate healing in rats and that males were benefited to a greater degree than females. In retrospect, it is likely that the commercial chow given to the rats was zinc deficient and that the animals were consequently zinc deficient. If today's routinely zinc-enriched feeds had been in use then, the observation could not have been made.

Zinc Tropism of Healing Tissues

If it is true that zinc promotes wound healing, it might be expected that zinc

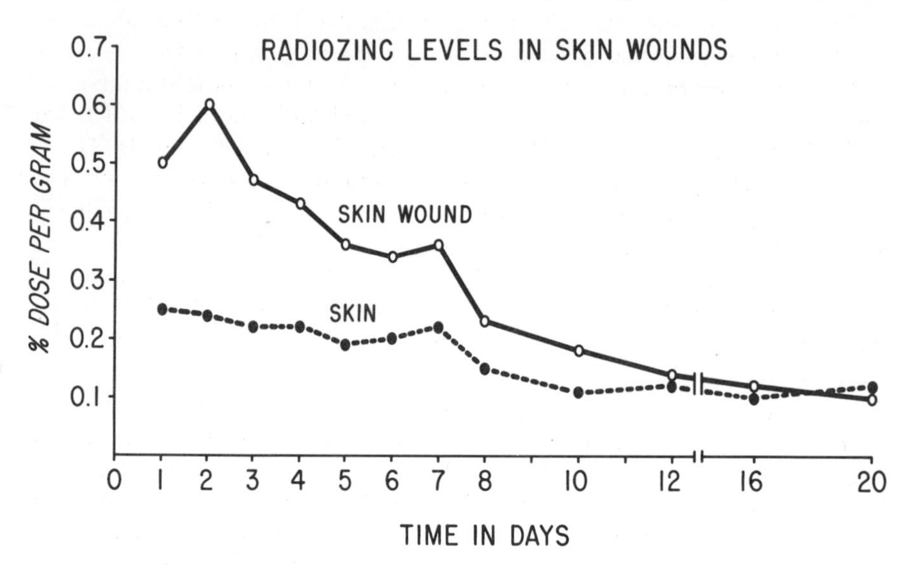

Figure 7A. Increased radiozinc retention in incised skin wounds of rats compared to normal skin following intravenous injection of zinc-65 chloride at the time of wounding. The radioisotope accumulates preferentially in the wound during the acute healing stage.

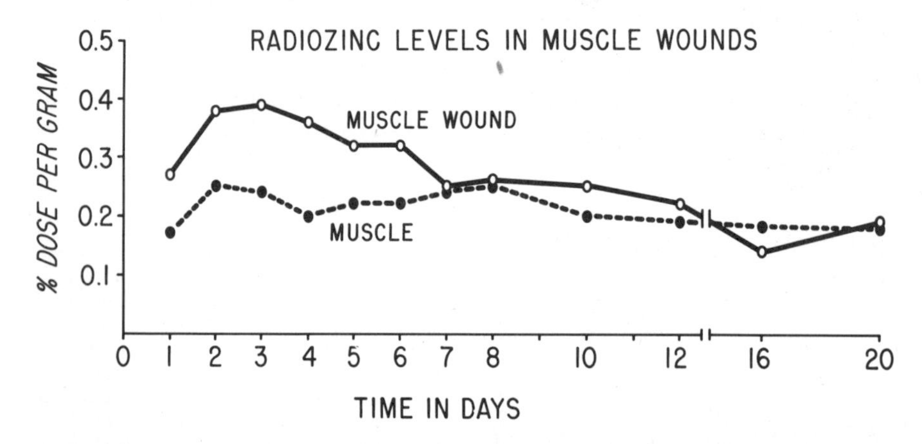

Figure 7B. Increased radiozinc retention in muscle wounds of rats compared to normal muscle following intravenous injection of zinc-65 chloride at the time of wounding. The radioisotope accumulates preferentially in the muscle wound during the acute healing stage, but the effect is not as pronounced as with the skin wound.

would accumulate in the healing tissues. To test this concept, we (Savlov and others, 1962) injected a series of rats with zinc-65 at the time of wounding, and then we compared the uptake of the radioisotope in the healing lesion with that in the adjacent nontraumatized tissues. As shown in Figures 7A and B, the radiozinc was preferentially concentrated in the healing tissues, with a peak of activity during the first 7 days after injury. Skin wounds showed a greater concentration of radioactivity than muscle wounds, when each was compared with the corresponding normal tissue. The migration of zinc-65 into the wounds was transient, as there was no evidence of radioactivity in the scar tissue 100 days after injury and injection. In contrast, the radioisotope was still detectable after this time in other tissues. These studies suggest that zinc is incorporated into the enzyme systems which are intimately involved in the healing process and that only a small fraction is involved in the structural linkages of collagen or the ground substance.

Zinc is also required for the repair of bones and arteries. Haumont and McLean (1966) showed that zinc accelerates the calcification of preosseous tissue and, subsequently, becomes incorporated into the mineral matrix of growing and healing bone. Studies in progress on the uptake of zinc-65 by healing arterial walls show that the radioisotope concentrates in the wound during the first week of repair.

Further evidence that adequate zinc is important in healing has been gained from observations in animals that were fed diets inadequate in zinc. Zinc-deficient cattle, poultry, sheep, and swine spontaneously develop ulcerating lesions in areas of trauma, such as the legs, which heal only if adequate dietary zinc is restored. Miller and others (1965) noted the dependency of healing on adequate zinc stores and suggested that poor healing in cattle can be used as a sign of zinc deficiency. Sandstead and Shepard (1968) demonstrated the decreased tensile strength of wounds in zinc-deficient rats 14 days after injury.

Zinc Deficiency in Burned Patients

The healing wound places great demands on the zinc stores in the body. This became very apparent during our (Pories and others, 1967a) studies of patients with major burns. Figure 8 shows a composite graph of the changes in the hair zinc levels of a group of 47 burned patients at the Brooke Army Burn Center. The consistency of the curves is striking. All the patients had virtually identical patterns, which consisted of a rapid initial drop in the zinc levels that only began to return to normal at 90 to 120 days. These studies suggest that patients with major burns develop serious deficits and that zinc deficiency may be one of the causes of the difficult healing problems seen so often in these patients (Nielsen and Jemec, 1968).

The burned patient can develop zinc deficits for a variety of reasons: (1) The feedings, consisting of intravenous fluids or scanty meals, contain little zinc. (2) A major burn sequesters large amounts of zinc in the eschar; 20 percent of the body's zinc stores are in the skin. (3) The loss of skin is followed by a massive loss of serum proteins, which serve as carriers for the zinc ion.

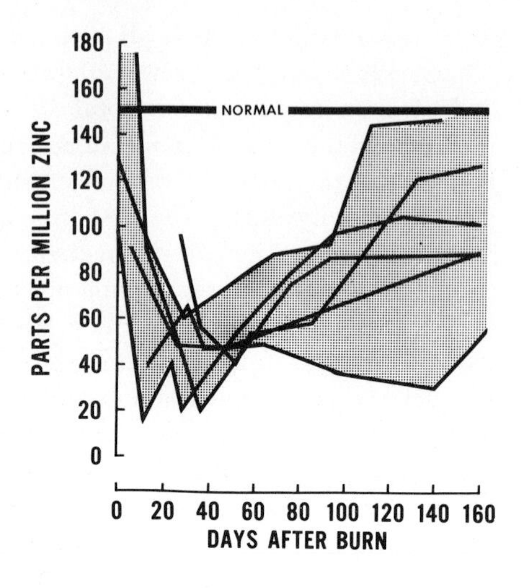

Figure 8. Changes in the zinc levels of hair after thermal burns. The dark lines show the changes in the zinc values with time in six patients with 20 percent or more third-degree burns. The cross-hatched areas between the extremes of these values represent the levels obtained in 47 patients with second- and third-degree burns.

(4) The burn is usually complicated by an infection; this results in a large loss of leukocytes which probably contain more zinc than any other cell in the body. (5) Some antibiotics are powerful chelating agents which may lead to zinc losses by binding the ion prior to excretion of the drug.

Acceleration of Human Healing with Zinc Sulfate

In 1965 we (Pories and others, 1967) designed a study to determine whether the addition of easily available zinc to the diet could accelerate the rate of the wound healing in humans. Twenty young airmen with pilonidal sinuses were randomized into two groups under statistical control prior to excision of the sinuses. Ten of the patients were placed on zinc sulfate USP ($ZnSO_4 \cdot 7H_2O$), 220 mg three times a day, while the ten in the control group were maintained on their usual diets without zinc supplementation. The methods of excision, dressing care, and activity were comparable in the two groups. The wounds were measured every five days with Jeltrate impressions (Fig. 9), and the final day of healing for each patient was recorded. The difference was dramatic. Although the medicated patients had larger wounds than the controls (54.5 ml *vs.* 32.3 ml), the airmen treated with zinc healed 34.3 days faster (48.8 ± S.E. 2.6 days *vs.* 80.1 ± S.E. 13.7 days; $p < 0.02$). This study demonstrated that zinc plays a role in human healing and that the administration of this element can benefit healing in humans. It also disturbingly suggests that the diet of these airmen may well have been zinc deficient or that even apparently healthy young men may not be able to mobilize adequate zinc stores to meet the demands of a healing wound.

Zinc Deficiency as a Cause of Delayed Wound Healing

That zinc deficiency is a common cause of delayed healing in hospital patients has been brought out in our recent study of 17 patients hospitalized with chronic, indolent wounds (Pories and others, 1969). The patients were referred to us because their wounds failed to heal over a long period of time despite good wound care and nutrition on balanced hospital diets. To avoid bias, all 17 patients referred to us were accepted for study, although only 12

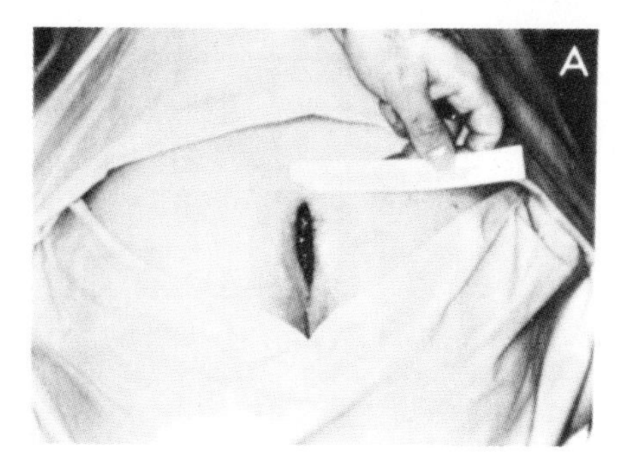

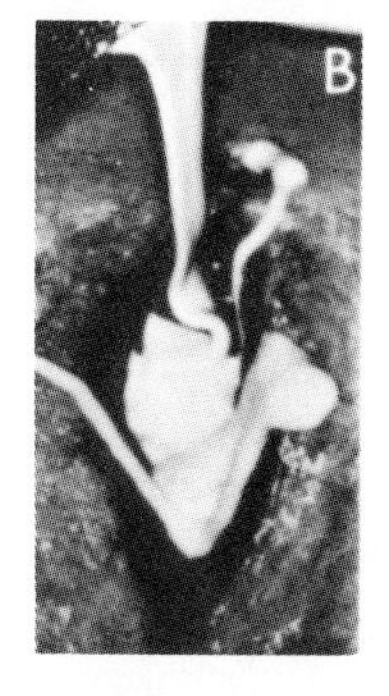

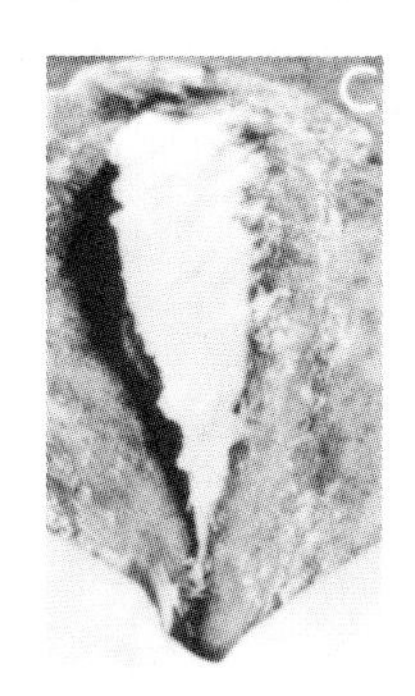

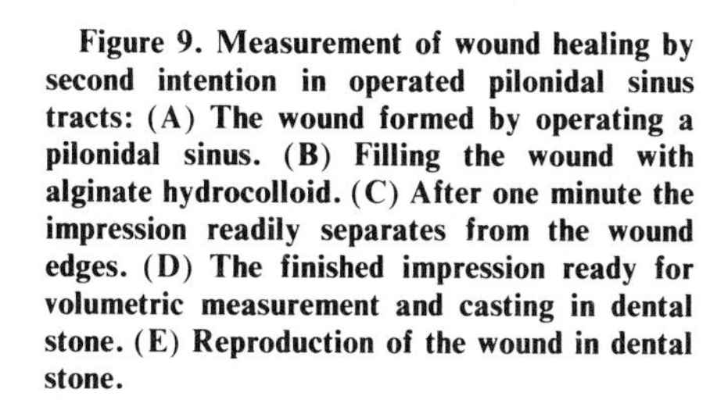

Figure 9. Measurement of wound healing by second intention in operated pilonidal sinus tracts: (A) The wound formed by operating a pilonidal sinus. (B) Filling the wound with alginate hydrocolloid. (C) After one minute the impression readily separates from the wound edges. (D) The finished impression ready for volumetric measurement and casting in dental stone. (E) Reproduction of the wound in dental stone.

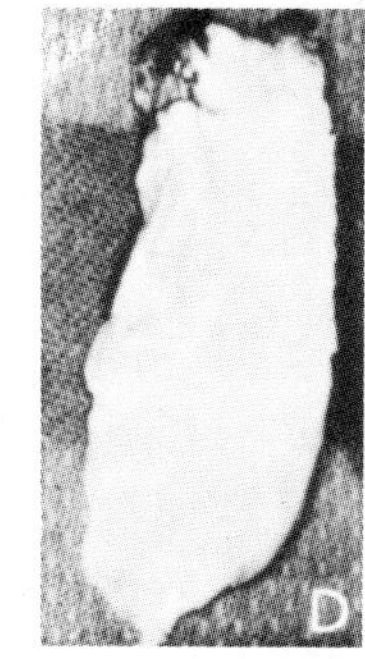

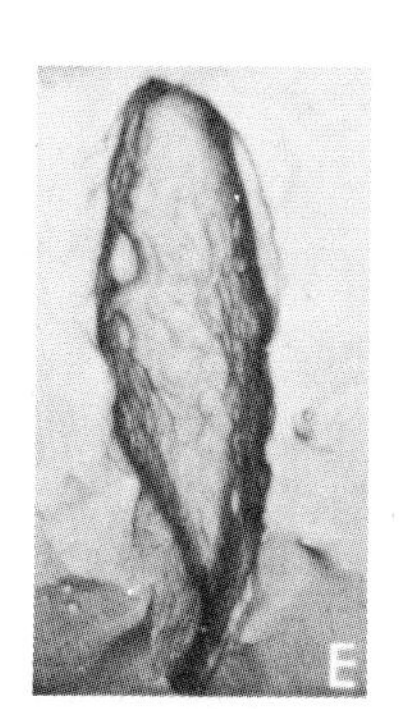

of the 17 were zinc deficient. All the patients were treated with oral supplements of zinc sulfate USP ($ZnSO_4 \cdot 7H_2O$), 220 mg in #2 gelatin capsules, administered three times daily with milk to minimize any gastric irritation. Serum zinc levels were determined by atomic absorption spectroscopy both before and during treatment. Local and systemic wound therapy was continued without change. The progress of the healing was followed by clinical impressions and photographs.

The results of zinc sulfate therapy in the 17 patients with indolent wounds are summarized in Table 1. All the patients with zinc levels below 110 mcg percent responded to zinc therapy; the others did not. The relationship of pretreatment serum zinc levels to the effect of zinc supplementation is illustrated in Figure 10. The mean serum zinc level of the responding patients was 75.1 mcg percent *versus* 125.2 mcg percent in those failing to respond. There was no overlap between the two groups. Statistical analysis showed that the zinc sulfate therapy promoted healing in the zinc-deficient patients. With a t-test of the pooled variances with 15° of freedom, the result was p ($t = 5.34$) <0.005.

The response to zinc therapy was definite in the zinc-deficient patients; epithelial ingrowth resumed rapidly, granulation turned the bases a brilliant red, and wound drainage decreased quickly. In some cases, these changes were apparent within 4 days after beginning zinc therapy, although the serum zinc levels usually required 4 to 6 weeks to return to normal. Zinc sulfate

Table 1. Zinc Sulfate Therapy in Patients with Delayed Wound Healing

Patient	Age	Diabetes	Serum Zinc (mcg%) Pre-Rx	Serum Zinc (mcg%) Post-Rx	Problem	Result of Zinc Therapy
Zinc-Deficient Patients						
1. A.W.	46	+	84	169	Gangrenous amputation site	Granulated, grafted, healed
2. J.B.	25	0	44	179	Indolent thoracotomy wound	Healed
3. G.S.	56	+	78	110	Gangrenous amputation site	Granulated, grafted, healed
4. W.H.	62	0	54	156	Chronic decubitus	Granulated, grafted, healed
5. E.S.	63	+	90	191	Gangrenous amputation site	Healed
6. R.C.	59	0	43	84	Multiple chronic decubiti	Healed
7. F.P.	84	+	86	310	Malleolar ulcer	Healed
8. L.C.	66	0	77	137	Amputation stump ulcer	Healed
9. B.M.	65	0	92	142	Chest wall sinus	Healed
10. M.L.	64	0	82	141	Gangrenous amputation site	Granulated, grafted, healed
11. T.Y.	61	+	68	148	Gangrenous metatarsal ulcer	Healed
12. S.B.	35	0	103	184	Chronic decubitus	Healed
Zinc-Sufficient Patients						
13. L.K.	62	0	152	—	Gangrenous amputation site	No change, died
14. A.H.	53	0	116	—	Chronic sinus	No change
15. M.H.	84	0	113	180	Chronic heel ulcer	No change
16. P.D.	77	+	134	—	Gangrenous metatarsal ulcer	No change, died
17. F.C.	65	0	111	181	Gangrenous metatarsal ulcer	No change, amputated

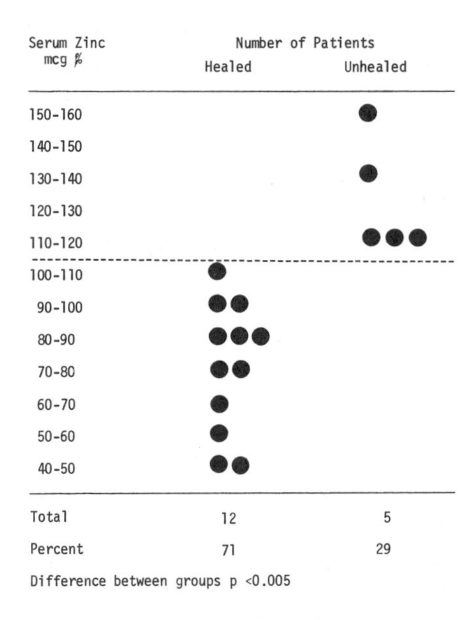

Figure 10. Correlation of initial serum zinc levels with healing response of 17 patients to zinc sulfate therapy.

therapy was well tolerated, even over long periods of time. Several representative case histories are presented.

(1) A. W.: A 46-year-old woman with a history of brittle diabetes of 30-years duration had undergone transmetatarsal amputation on January 1, 1967, following an unsuccessful femoropopliteal vein graft for peripheral gangrene. The wound not only failed to heal but also soon developed advancing gangrenous edges. The patient refused higher amputation and was starting on zinc sulfate therapy on January 11, 1967. The lowest serum zinc value was 84 mcg percent. Within four days the granulations became bright red and an advancing edge of epithelium became evident. A skin graft placed on January 27, 1967 healed completely, and she was discharged on February 18, 1967. She continued to ambulate on the foot without difficulty.

(2) W.H.: A 62-year-old worker with flaccid paraplegia below the T-10 level had multiple unsuccessful operations for the closure of sacral decubiti. A large ulcer entending from S-2 to the scrotum, with a vesicosacral fistula was present since May 1967. It had failed to respond to repeated debridement and meticulous dressing care during a 9-week period on the plastic surgical services. His lowest serum zinc level was 54 mcg percent; on 7/23/68, he was treated with zinc sulfate. The wound response was remarkable, and a healthy base with advancing granulations was evident in 5 days. On August 16, 1968, the ulcer was successfully covered with a skin graft. His back has remained healed.

(3) E.S.: A 63-year-old diabetic laborer had undergone open amputation of his right great toe and the first metatarsal on March 8, 1968, after repeated debridement of the advancing gangrene had proven unsuccessful. Over the next 8 days, the gangrene continued to progress and below knee amputation was advised by several consultants. On March 16, 1968, a trial of zinc sulfate therapy was begun. His serum zinc level was 90 mcg percent. Within 4 days, the wound was clean and bright red, and by 4/17/68 healing was complete. He has continued to walk on the foot without difficulty.

These studies support the thesis that zinc deficiency is a common cause of delayed healing in man. Correction of the deficit with zinc sulfate is safe and is associated with the resumption of normal healing in zinc-deficient patients.[2] As might be expected, the administration of zinc supplements to zinc-sufficient patients is not harmful, but it has no effects on delayed healing due to other causes.

ZINC AND CHRONIC DISEASE

Atherosclerosis not only is the most common cause of death in the United States, but also is one of the most vexing chronic diseases in medical research because it appears to be partly an environmental and preventable disease. A number of causes have been implicated, but to date alteration of these factors has not led to any significant decrease in the incidence of the disease.

Zinc and Atherosclerosis

It has recently become apparent that mineral imbalances are additional factors in arterial disease. Deficiencies of zinc, copper, and vanadium may interfere with arterial metabolism; excesses of cadmium, calcium, and cobalt may be injurious to the arterial wall.

The first clue that atherosclerosis was related to the deficiency of zinc came during a trace-metal survey of a number of patients with different diseases. Patients with atherosclerosis uniformly demonstrated low zinc values. Hair zinc levels were examined in 25 unselected male patients who were admitted to the University of Rochester Medical Center with the diagnosis of atherosclerosis. The diagnosis was proven in all cases by arteriography and usually by surgery. In contrast to the normal hair levels of 125 to 150 ppm (Pories and others, 1966), these patients had a mean of 62.2 ppm, or less than half of normal, as shown in Table 2. In this table are given, also, the effects of zinc therapy on hair and blood zinc levels.

The finding that patients with arterial disease have low zinc levels has been confirmed by two other independent investigators. Vallee (Vallee and others, 1956) of Harvard reported that patients with myocardial infarctions have serum zinc levels of 67 ppm/100 ml compared to normal levels of 120/100 ml. Volkov (1963) of the Minsk Medical Institute of Russia also studied zinc levels in 72 patients with atherosclerosis and found similarly low values.

The roles played by elements in the arterial wall are not known, but an element which favors healing seems to have a place in checking the develop-

[2]The controlled trial of oral zinc sulfate therapy by Husain (1969) is an elegant demonstration of how zinc medication should be used. One hundred and four inpatients of various ages and both sexes with ulcers of the lower limb were studied. Identical capsules containing 220 mg of lactose (placebo) or zinc sulphate were given three times a day to equal numbers of controls and treated patients. Ulcers healed more quickly in the treatment group. The ulcers of the treatment group healed in 32 days on average, as against 77 days for the control group. There was no significant dependence on age or sex. The mean serum zinc level in a pilot study was μg/ml ± S.E. 1.015 (±0.0406) μg per ml at the first estimation, 1.886 (±0.2405) μg per ml during treatment, and μg/ml 1.282 (±0.1149) μg per ml after treatment. The average surface area of the ulcers was 4290 sq mm in the control group and 4830 sq mm in the treatment group, and the healing rates were 90 sq mm per day and 421.8 sq mm per day, respectively.

Table 2. Zinc Therapy and Changes in Hair and Serum Zinc Levels

Medical status	Zinc Level ± S. E. Hair (ppm)	Serum (mcg%)
Normals	125–150	120 ± 19
Dwarfs	54.1 ± 5.5	73 ± 6*
Zn-treated dwarfs	121.1 ± 4.8	—
Atherosclerotics	62.2 ± 4.2	85 ± 15
Zn-treated atherosclerotics	—	167.2 ± 14
Healing problems	—	75.2 ± 2.8
Zn-treated Healing Problems	—	162.5 ± 5.1

*Plasma

ment of the pathology in the blood vessels. Duff (1954) pointed out in an excellent review that "Every experimental observation that has ever been recorded bearing on the effect of injury on the arteries had indicated that damage to the arterial walls has a localizing and promoting influence on the development of atherosclerosis, provided that appropriate systemic conditions are present." More recently, other pathologists (Chapman, 1965; Constantinides, 1966; Friedman and Van den Bovenkamp, 1966) have emphasized the initiation of arterial damage through failure of the blood vessel to heal properly after trauma. Delayed healing in the arteries probably predisposes to necrosis, intimal proliferation, and cholesterol deposition. Our observation that radiozinc is preferentially concentrated in healing tissues gives further support to the thesis that zinc is required for arterial repair.

More than three years ago we began a program of zinc medication in patients with advanced inoperable atherosclerosis. Thirty-six patients with severe, incapacitating disease were treated with zinc sulfate USP, 220 mg three times a day for periods up to 39 months. All tolerated the medication well. Thirty patients (83 percent) improved symptomatically with increases of exercise tolerance and leg warmth. Previously absent pulses returned in 14 (39 percent) of the 36 pulses. Arteriograms repeated after one year showed no change in the major vessels, but suggested an increase in the collateral channels. Five of six patients healed gangrenous ischemic lesions. During the study two patients died from pre-existing heart failure, reflecting the marginal medical status of some of these patients. Those who benefited by the zinc therapy had lower serum zinc levels than patients not helped by the treatment (85 ± 15 *vs.* 140 ± 38 mcg percent, $p < 0.005$). Henzel and his associates (1969) at the University of Missouri recently reported similar data in 12 patients. These results are very encouraging inasmuch as atherosclerosis is a disease which generally becomes progressively worse and rarely shows spontaneous improvement.

Zinc and Chronic Disease

Chronic disease is generally the result of continued injury or of poor tissue repair. The infected wound, the indolent decubitus ulcer, many skin diseases, and the destructive scarring of Laennec's cirrhosis, rheumatic valvulitis, bronchiectasis, and rheumatoid arthritis are all manifestations of disordered healing. The prevention and treatment of chronic disease are the main objectives of medicine, which is why so much medical research has been directed toward a better understanding of the healing process. With evidence that zinc is also required for normal healing, the deficiency of this element has become a central interest in much current medical research. The association of low zinc levels with chronic disease, such as cirrhosis (Vallee and others, 1956; Sullivan and others, 1968), lung cancer (Davies and others, 1968), myocardial infarction (Volkov, 1963) and certain hematological disorders (Prasad, 1966), is a provocative lead that needs to be thoroughly investigated. If zinc deficiency does play a role in the pathogenesis of chronic disease, correction with zinc sulfate would be a simple, safe, and effective therapy.

SUMMARY

Zinc deficiency is common in crops, domestic animals, and man. Therapy with zinc preparations has resulted in spectacular gains in yields by plants. Similarly, addition of supplemental zinc to animal feeds has given substantial gains in health and feed efficiency, particularly with poultry and swine. Zinc deficiency is now recognized as a common, widespread nutritional disorder in agriculture. A great deal more must be done to correct this deficiency.

In man, zinc deficiency interferes with development, growth, fertility, and tissue repair. Oral zinc sulfate medication has promoted growth in nutritional dwarfs, has accelerated healing in young men and in elderly patients with indolent wounds, and has improved the ischemia of peripheral atherosclerosis. Zinc therapy and its beneficial effects on tissue repair must be studied in many other chronic disease states as zinc deficiency appears to be so common in hospital patients.

ACKNOWLEDGMENTS

This research was supported in part by U.S. Public Health Service Grant HE 10213, National Institutes of Health; Research Grant RH 00042, National Center of Radiological Health; the Horatio H. Burtt and Clarence S. Lunt Research Funds of the University of Rochester; and a grant from the Aerospace Medical Division, under approval of the Surgeon General, U.S. Air Force. The contents of the paper reflect the authors' personal views and are not to be construed as a statement of official Air Force policy. The animal experiments were conducted according to the "Rules Regarding Animal Care," established by the American Medical Association.

REFERENCES CITED

Anonymous, 1964, Approximate composition and nutritive values of dry milks: Chicago, American Dry Milk Institute, 4 p.

Archibald, J. G., 1958, Trace elements in milk: A review, part II: Dairy Sci. Abs., 20, p. 799–812.

Berger, K. C., 1962, Micronutrient deficiencies in the United States: Agr. and Food Chemistry, v. 10, p. 178–181.

Berry, R. K., Bell, M. C., and Wright, P. L., 1966, Influence of dietary calcium, zinc, and oil upon the in vitro uptake of zinc-65 by porcine blood cells: Jour. Nutrition, v. 88, p. 284–290.

Blamberg, D. L., Blackwood, U. B., Supplee, W. C., and Combs, G. F., 1960, Effect of zinc deficiency in hens on hatchability and embryonic development: Soc. Exp. Biol. Medicine Proc., v. 104, p. 217–220.

Caggiano, V., Schnitzler, R., Strauss, W., Baker, R. K., Carter, A. C., Josephson, A. S., and Wallach, S., 1969, Zinc deficiency in a patient with retarded growth, hypogonadism, hypogammaglobulinemia and chronic infection: Am. Jour. Med. Sciences, v. 257, p. 305–319.

Carter, J. P., Grivetti, L. E., Davis, J. T., Nasiff, S., Mansour, A., Mousa, W. A., Atta, Alaa-El-Din, Patwardhan, V. N., Moneim, M. A., Adbou, I. A., and Darby, W. J., 1969, Growth and sexual development of adolescent Egyptian village boys: Am. Jour. Clin. Nutrition, v. 22, p. 59–78.

Chapman, I., 1965, Morphogenesis of occluding coronary artery thrombosis: Arch. Pathology, v. 80, p. 256–261.

Coble, Y. D., Van Reen, R., Schulert, A. R., Koshakji, R. P., Farid, Z. and Davis, J. T., 1966, Zinc levels and blood enzyme activities in Egyptian male subjects with retarded growth and sexual development: Am. Jour. Clin. Nutrition, v. 19, p. 415–421.

Constantinides, P., 1966, Plaque fissures in human coronary thrombosis: Jour. Atherosclerosis Research, v. 6, p. 1–17.

Davies, I. J. T., Musa, M., and Dormandy, T. L., 1968, Measurements of plasma zinc: Jour. Clin. Pathology, v. 21, p. 359–365.

Duff, G. L., 1954, Summary of part II. Symposium on atherosclerosis: Washington, D.C., National Academy of Sciences, National Research Council, Pub. 338, 127 p.

Friedman, M., and Van den Bovenkamp, G. J., 1966, The pathogenesis of a coronary thrombus: Am. Jour. Pathology, v. 48, p. 19–44.

Haumont, S., and McLean, F. C., 1966, Zinc and the physiology of bone, *in* Prasad, A. S., *Editor*, Zinc Metabolism: Springfield, Illinois, Charles C Thomas, p. 169–186.

Henzel, J. H., Holtman, B., Keitzer, F. W., De Weese, M. S., and Lichti, E., 1969, Trace elements in atherosclerosis, efficacy of zinc medication as a therapeutic modality: 2nd Ann. Missouri Conf. on Trace Substances in Environmental Health Proc., p. 83–99.

Hoekstra, W. G., Lewis, P. K., Jr., Phillips, P. H., and Grummer, R. H., 1956, The relationship of parakeratosis, supplemental calcium and zinc to the zinc content of certain body components of swine: Jour. Animal Sci., v. 15, p. 752–764.

Hurley, L. S., and Swenerton, H., 1966, Congenital malformations resulting from zinc deficiency in rats: Soc. Exp. Biol. Medicine Proc., v. 123, p. 692–696.

Husain, S. L., 1969, Oral zinc sulphate in leg ulcers: Lancet, v. I, p. 1069–1071.

Kienholz, E. W., Turk, D. E., Sunde, M. L., and Hoekstra, W. G., 1961, Effects of zinc deficiency in the diets of hens: Jour. Nutrition, v. 75, p. 211–221.

Miller, J. K., and Miller, W. J., 1960, Development of zinc deficiency in Holstein calves fed a purified diet: Jour. Dairy Sci., v. 43, p. 1854–1856.

Miller, W. J., Blackmon, D. M., Gentry, R. P., Powell, G. W., and Perkins, H. F., 1966, Influence of zinc deficiency on zinc and dry matter content of ruminant tissues and on excretion of zinc: Jour. Dairy Sci., v. 49, p. 1446–1453.

Miller, W. J., Morton, J. D., Pitts, W. J., and Clifton, C. M., 1965, Effects of zinc deficiency and restricted feeding on wound healing in the bovine: Soc. Exp. Biol. Medicine Proc., v. 118, p. 427–430.

Nielsen, S. P. and Jemec, B., 1968, Zinc metabolism in patients with severe burns: Scandinavian Jour. Plast. Reconstr. Surgery, v. 2, p. 47–52.

O'Dell, B. L., and Savage, J. E., 1957, Symptoms of zinc deficiency in the chick: Fed. Proc., 16:394.

Pories, W. J., Henzel, J. H., Rob, C. G., and Strain, W. H., 1967a, Acceleration of healing with zinc sulfate: Surgery Ann., v. 165, p. 432–436.

Pories, W. J., Rob, C. G., Smith, J. L., Henzel, J. H., and Strain, W. H., 1966, Zinc deficiency, another cause of atherosclerosis: 7th Internat. Cong. Gerontology Proc., p. 449.

Pories, W. J., Strain, W. H., Peer, R. M., and Landew, M. H., 1969, Zinc deficiency as a cause for delayed wound healing: Current Topics in Surg. Research, v. 1, p. 315–323.

Pories, W. J., Strain, W. H., Rob, C. G., Henzel, J. H., Hennessen, J. A., and Plecha, F. R., 1967, Trace elements and wound healing: 1st Ann. Missouri Conf. on Trace Substances in Environmental Health Proc., p. 114–133.

Prasad, A. S., 1966, Metabolism of zinc and its deficiency in human subjects, *in* Prasad, A. S., *Editor,* Zinc Metabolism: Springfield, Illinois, Charles C Thomas, p. 250–304.

Prasad, A. S., Miale, A., Jr., Farid, Z., Sanstead, H. H., Schulert, A. R., and Darby, W. J., 1963, Biochemical studies on dwarfism, hypogonadism and anemia: Am. Med. Assoc. Arch. Int. Medicine, v. 111, p. 407–428.

Price, C. A., 1963, Control of processes sensitive to zinc in plants and microorganisms; *in* Prasad, A. S., *Editor,* Zinc Metabolism: Springfield, Illinois, Charles C Thomas, p. 69–89.

Ronaghy, H., Spivey Fox, M. R., Garn, S. M., Israel, H., Harp, A., Moe, P. G. and Halsted, J. A., 1969, Controlled zinc supplementation for malnourished school boys: a pilot experiment: Am. Jour. Clin. Nutrition, v. 22, p. 1279–1289.

Sandstead, H. H., and Shepard, G. H., 1968, The effect of zinc deficiency on the tensile strength of healing surgical incisions in the integument of the rat: Soc. Exp. Biol. Medicine Proc., v. 128, p. 687–689.

Savlov, E. D., Strain, W. H., and Huegin, F., 1962, Radiozinc studies in experimental wound healing: Jour. Surg. Research, v. 2, p. 209–212.

Schneider, E., and Price, C. A., 1962, Decreased ribonucleic acid levels: A possible cause of growth inhibition in zinc deficiency: Biochim. et Biophys. Acta, v. 55, p. 406–407.

Stiles, W., 1961, Trace Elements in Plants, 3rd ed.: Cambridge, Cambridge Univ. Press, 230 p.

Strain, W. H., Dutton, A. M., Heyer, H. B., and Ramsey, G. H., 1953, Experimental studies on the acceleration of burn and wound healing: Rochester, New York, Rochester Univ. Rep., 18 p.

Strain, W. H., Lascari, A., and Pories, W. J., 1966, Zinc deficiency in babies: 7th Internat. Cong. Nutrition Proc., v. 5, p. 759–763.

Sullivan, J. F., Parker, M. M., and Boyett, J. D., 1968, Incidence of low serum zinc in noncirrhotic patients: Soc. Exp. Biol. Medicine Proc., v. 130, p. 591–594.

Swenerton, H., and Hurley, L. S., 1968, Severe zinc deficiency in male and female rats: Jour. Nutrition, v. 95, p. 8–18.

Todd, W. R., Elvehjem, C. A., and Hart, E. B., 1934, Zinc in the nutrition of the rat: Am. Jour. Physiology, v. 107, p. 146–156.

Tomita, K., 1961, Influence of thiamine on the metabolism of zinc: I. Determinations of zinc: Bitamin (Kyoto), v. 22, p. 31–34.

Tsui, C., 1948, The role of zinc in auxin synthesis in the tomato plant: Am. Jour. Botany, v. 35, p. 172–179.

Tucker, H. F., and Salmon, W. D., 1955, Parakeratosis or zinc deficiency disease in the pig: Soc. Exp. Biol. Medicine Proc., v. 88, p. 613–616.

Underwood, E. J., 1962, Trace Elements in Human and Animal Nutrition, 2nd ed.: New York, Academic Press, 429 p.

Vallee, B. L., Wacker, W. E. C., Bartholomay, A. F., and Robin, E. D., 1956, Zinc metabolism in hepatic dysfunction I. Serum zinc concentrations in Laennec's cirrhosis and their validation by sequential analysis: New England Jour. Medicine, v. 255, p. 403–408.

Viets, F. G., Jr., 1963, Zinc deficiency in the soil-plant system; *in* Prasad, A. S., *Editor,* Zinc Metabolism: Springfield, Illinois, Charles C Thomas, p. 90–128.

Volkov, N F., 1963, Cobalt, manganese and zinc content in the blood of atherosclerosis patients: Fed. Proc. (Transl. Suppl.), v. 22, p. T897–899.

Wacker, W. E. C., 1962, Nucleic acids and metals III. Changes in nucleic acid, protein, and metal content as a consequence of zinc deficiency in *Euglena gracilis:* Biochemistry, v. 1, p. 859–865.

Wacker, W. E., Ulmer, D. D., and Vallee, B. L., 1956, Metalloenzymes and myocardial infarction: II. Malic and lactic dehydrogenase activities and zinc concentrations in serum: New England Jour. Medicine, v. 255, p. 449–456.

Zeigler, T. R., Scott, M. L., McEvoy, R. K., Greenlaw, R. H., Huegin, F., and Strain, W. H., 1962, Radiographic studies on skeletal parts of zinc deficient pullets: Soc. Exp. Biol. Medicine Proc., v. 109, p. 239–242.

PRESENT ADDRESS (PORIES AND STRAIN): DEPARTMENT OF SURGERY, CLEVELAND METROPOLITAN GENERAL HOSPITAL, 3395 SCRANTON ROAD, CLEVELAND, OHIO 44109

MANUSCRIPT RECEIVED BY THE SOCIETY MAY 18, 1970

Printed in U.S.A.

THE GEOLOGICAL SOCIETY OF AMERICA, INC.
MEMOIR 123, 1971

Variations in the Copper, Zinc, Lead, and Molybdenum Contents of Some Vegetables and Their Supporting Soils

HARRY V. WARREN
AND
ROBERT E. DELAVAULT
Department of Geology, University of British Columbia
Vancouver, British Columbia, Canada

ABSTRACT

Variations in the trace elements content of different vegetables are far greater than generally realized. These variations are caused by a number of factors, including geologic and geographic ones and industrial pollution.

Original data are presented to show the varying concentrations of copper, zinc, lead, and molybdenum in samples of the edible portions of lettuce, cabbage, potato, carrot, and bean plants collected from various districts in Great Britain and Canada.

Too few samples have been dealt with to establish mean values that are statistically significant, but on the basis of the evidence available, there is a distinct possibility that unusual patterns of disease distribution may be causally related to anomalous concentrations of one or more trace elements. Illustrative data on such a (causal?) association are presented.

CONTENTS

Table

INTRODUCTION

For more than 20 years the authors have been developing the means by which biogeochemistry may be used as a tool in the search for buried ore bodies. The fact that large variations occur in the trace-element content of various trees and lesser plants prompted the authors to find out whether or not there are comparable variations in the quantities of these elements occurring in some of the more commonly used vegetables.

We have found by experience that it is not always possible to predict the trace-element content of vegetal matter even when the *p*H, the soil fractions, as well as the organic and trace-element content of the parent soil are known. Consequently, in the hope that our data would be sufficiently interesting to justify further studies, we have investigated the relationships between the trace-element content of various vegetables and the soils on which they were grown. In accordance with our normal laboratory practice, our soil contents are reported in terms of sulphuric-acid extractable metal (Warren and Delavault, 1956). Plant material was ashed at 550° C for 12 hours and chemically analyzed. Recently, copper and zinc determinations were checked by atomic absorption. Thus the trace-element content of the soils on which we are reporting will not be strictly comparable with what would be reported by other workers who seem to be more in favor of determining either the total metal content of a soil or of reporting the much smaller fraction of the total metal which is obtained by using a relatively weaker extractant.

From our own experience with many thousands of analyses, we believe that, in general, correlations between the trace-element content of soils and vegetables growing on them are better observed when the soil content is reported in terms of sulphuric-acid extractable metal rather than as total metal or as that fraction available to some weaker extractant. Our experience would suggest that no single method of reporting the metal content of a soil is ideal and probably when many further investigations have been made, it may be found that "tailor-made" techniques for dealing with specific problems should be used. However, the important point surely is the total metal content of a vegetable; in this respect our results should be comparable with those of other workers.

In Tables 1 through 5 we hope to make clear how the trace-element content of various vegetables varies more widely than generally has been realized and how the trace-element relationship between soil and vegetable is more complex than might be anticipated. We have omitted references to soil *p*H, to the organic content of the soils, and to the results obtained by using other extractants, in order to keep the tables simple. We are sure that no other selection of data would have materially altered our general conclusions.

METHODS OF COLLECTING SAMPLES

Limited financial resources prevented us from collecting all our own samples. The late Dr. Allen-Price and Miss P. Fletcher were responsible for most of the Devonshire material, Dr. Eric Wilkes for a useful collection from Derbyshire, and Dr. N. Mitchel and Dr. R. S. Allison for those samples from Northern Ireland and from Orkney and Shetland Islands, respectively. Warren made an extensive collection of soils and vegetal matter in England, but by far the most diverse collection of soils and vegetables was sent to us by various volunteer members of the Royal College of General Practitioners acting under the general guidance of Dr. R. J. F. H. Pinsent and Dr. D. L. Crombie, who, as Research Advisor and Director, respectively, of that body, have sparked enthusiasm for exploratory research in the field of trace elements and epidemiology. Dr. W. G. Smitheringale of the Department of Geology at Memorial University, Newfoundland, was kind enough to make available to us soils and vegetables from several different areas of Newfoundland.

The Point Grey and Fraser Valley soils from British Columbia originate from fluvial and glacial parent material and have developed in a marine West-Coast climate; the soils from Nelson and Riondel in the Kootenay District and those from Peachland in the Okanagan District also have a fluvioglacial origin, but have developed in a continental climate. The British samples originate from a wide variety of parent material and a broad range of climate, ranging from a near-Mediterranean climate in southwest Cornwall to a relatively harsh marine climate in Shetland and Orkney Islands.

From the above remarks it should be clear that we are *not* attempting to say what are the "normal" or "average" amounts of any element in any vegetable. We are attempting primarily to demonstrate the range of concentrations that may occur in the British Isles and British Columbia in areas where unusual epidemiological abnormalities are not generally recognized.

The Riondel samples might have been excluded. They come from soils demonstrably high in lead and zinc. However, gardens there are particularly cherished, and the vegetables and the inhabitants alike seem to flourish. The Riondel "Z" samples were from a garden rich in organic material; the "S" samples were from a garden deficient in organic matter.

No significance should be read into the samples that illustrate the range of concentrations in each vegetable. Cabbage samples were abundant in Great Britain, but rare in British Columbia, and no lettuce samples came from Newfoundland.

TABLE 1. THE COPPER CONTENT OF VARIOUS VEGETABLES AND THEIR SUPPORTING SOILS (in ppm)

Country	County or Province	Parish or Locality		Dry Plant	Ash	Soil
		A. Lettuce (Edible Inner Leaves)				
England	Devonshire	Morwellham (1964)		15	150	140
Canada	British Columbia	Point Grey (1967)		13	173	21
England	Devonshire	Milton Abbot		13	110	40
Canada	British Columbia	Riondel (Z)		13	92	107
Canada	British Columbia	Fraser Valley		10	62	10
Canada	British Columbia	Point Grey (1968)		10	110	14
Canada	British Columbia	Nelson		10	38	19
England	Middlesex	Feltham		8	46	3
England	Isle of Man	Douglas		6	34	100
England	Kent	Ashford		3	32	20
			Range	3–15	32–173	
			Mean	10	85	
		B. Cabbage (Edible Inner Leaves)				
England	Worcestershire	Redditch		6	78	10
England	Yorkshire	South Milford		6	66	40
England	Cornwall	Probus		5	49	100
Scotland	Stirlingshire	Falkirk		5	58	40
England	Sussex	Petworth		4	48	16
England	Kent	Ashford		3	31	20
Canada	British Columbia	Riondel (Z)		2	11	107
England	Shropshire	Church Stretton		2	26	40
Scotland	Selkirkshire	Galashiels		2	22	48
England	Essex	Chelmsford		1	7	40
			Range	1–6	7–78	
			Mean	4	40	
		C. Potato (Tuber Peeled)				
Scotland	Stirlingshire	Falkirk		6	200	40
Canada	British Columbia	Riondel (Z)		6	160	107
Canada	Newfoundland	Winterland (S)		6	150	1.2
England	Derbyshire	Steetley		4	120	9
England	Nottinghamshire	Skegby		4	120	4
Wales	Cardiganshire	Aberystwyth		3	77	70
England	Shropshire	Church Stretton		2	41	40
Canada	British Columbia	Fraser Valley		2	41	12
England	Essex	Chelmsford		.7	17	40
England	Worcestershire	Redditch		.5	16	10
			Range	.5–6	16–200	
			Mean	3	94	
		D. Carrot (Edible Portion)				
Canada	Newfoundland	Fortune (P)		5	54	2
Canada	British Columbia	Riondel (S)		5	66	90
Canada	British Columbia	Nelson		5	57	19
Scotland	Berwickshire	Coldstream		4	80	5
England	Cornwall	Feock		3	57	80
Canada	British Columbia	Peachland		3	44	23
Canada	Newfoundland	Fortune (L)		3	27	54
England	Devonshire	Holsworthy		2	35	40
England	Cambridgeshire	Cambridge		2	32	32
Scotland	Dumfrieshire	Lynhurst		.6	8	20
			Range	.6–5	8–80	
			Mean	3	46	

TABLE 1. (CONTINUED)

E. Beans (Edible Portion—Various Varieties)						
Canada	British Columbia	Fraser Valley		9	120	12
Canada	British Columbia	Nelson		8	150	19
Canada	British Columbia	Point Grey		6	140	7
Canada	British Columbia	Peachland		6	110	23
Canada	British Columbia	Riondel (Z)		6	99	107
England	Devonshire	Morwellham		5	140	*
England	Northumberland	Hexham		4	90	2
England	Devonshire	Milton Abbot		4	50	*
Canada	British Columbia	Riondel (S)		4	41	91
England	Somerset	Axbridge		3	25	14
			Range	3–9	25–150	
			Mean	5.5	96	

*not available

TABLE 2. THE ZINC CONTENT OF VARIOUS VEGETABLES AND THEIR SUPPORTING SOILS (in ppm)

Country	County or Province	Parish or Locality		Dry Plant	Ash	Soil
A. Lettuce (Edible Inner Leaves)						
Canada	British Columbia	Riondel (S)		250	1100	1520
Canada	British Columbia	Riondel (Z)		150	980	394
Canada	British Columbia	Nelson		80	310	153
England	Middlesex	Feltham		78	450	61
England	Devonshire	Tavistock		62	550	290
England	Isle of Man	Douglas		58	330	630
England	Devonshire	Morwellham		56	570	430
Canada	British Columbia	Point Grey		54	853	28
Canada	British Columbia	Fraser Valley		44	280	520
England	Kent	Ashford		30	340	77
			Range	30–250	280–1100	
			Mean	86	576	
B. Cabbage (Edible Inner Leaves)						
Scotland	Stirlingshire	Falkirk		84	1000	170
England	Worcestershire	Redditch		60	730	150
England	Suffolk	Aldeburgh		52	610	35
Canada	British Columbia	Riondel (Z)		42	190	394
Canada	Newfoundland	Winterland (K)		38	430	31
England	Sussex	Petworth		32	380	130
Wales	Cardiganshire	Aberystwyth		28	380	710
England	Essex	Chelmsford		25	140	260
Canada	Newfoundland	Fortune (P)		16	190	21
Canada	Newfoundland	Fortune (L)		10	150	46
			Range	10–84	140–1000	
			Mean	39	420	
C. Potato (Tuber Peeled)						
England	Nottinghamshire	Skegby		29	880	121
Canada	British Columbia	Fraser Valley		28	450	52
Scotland	Stirlingshire	Falkirk		27	810	170
Canada	British Columbia	Point Grey		22	607	31
England	Middlesex	Hounslow		21	430	75
Canada	Newfoundland	Fortune (L)		17	360	38
Canada	British Columbia	Riondel (S)		16	340	1520

TABLE 2. (CONTINUED)

		C. Potato (Tuber Peeled)—Cont.				
Canada	British Columbia	Nelson		10	240	153
Wales	Glamorganshire	Swansea		9	260	300
England	Suffolk	Aldeburgh		8	170	35
			Range	8–29	170–880	
			Mean	18	450	
		D. Carrot (Edible Portion)				
Canada	British Columbia	Riondel (S)		59	780	1520
Canada	Newfoundland	Fortune (L)		42	430	138
Scotland	Selkirkshire	Galashiels		36	420	220
England	Cambridgeshire	Cambridge		29	390	410
Canada	Newfoundland	Fortune (P)		29	320	17
Canada	British Columbia	Nelson		26	310	153
Scotland	Berwickshire	Coldstream		23	450	56
Scotland	Dumfrieshire	Lynhurst		20	250	190
England	Gloucestershire	Cirencester		15	250	140
Canada	British Columbia	Peachland		12	190	29
			Range	12–59	190–780	
			Mean	29	380	
		E. Beans (Edible Portions—Various Varieties)				
England	Devonshire	Milton Abbot		94	740	*
England	Somersetshire	Axbridge		64	540	216
England	Devonshire	Morwellham		56	570	*
Canada	British Columbia	Riondel (Z)		46	760	394
Canada	British Columbia	Point Grey (1968)		43	570	26
Canada	British Columbia	Peachland		37	620	29
Canada	British Columbia	Nelson		32	570	153
England	Northumberland	Hexham		31	720	31
Canada	British Columbia	Point Grey (1967)		30	650	29
Canada	British Columbia	Riondel (S)		30	310	1520
			Range	30–94	310–760	
			Mean	46	605	

*not available

TABLE 3. THE LEAD CONTENT OF VARIOUS VEGETABLES AND THEIR SUPPORTING SOILS (in ppm)

		A. Lettuce (Edible Inner Leaves)				
Country	County or Province	Parish or Locality		Dry Plant	Ash	Soil
Canada	British Columbia	Riondel (S)		56	260	500
Canada	British Columbia	Riondel (Z)		32	210	180
England	Devonshire	Morwellham		11	110	200
England	Devonshire	Milton Abbot		9	81	80
Canada	British Columbia	Nelson		5	18	30
Canada	British Columbia	Point Grey		3	46	3
Canada	British Columbia	Fraser Valley		3	21	.8
England	Isle of Man	Douglas		2	14	400
England	Middlesex	Feltham		1	8	23
England	Kent	Ashford		.3	3	40
			Range	.3–56	3–260	
			Mean	12	67	

TABLE 3. (CONTINUED)

		B. Cabbage (Edible Inner Leaves)				
England	Warwickshire	Coleshill		2.3	24	110
England	Worcestershire	Redditch		1.8	22	48
Wales	Glamorganshire	Swansea		1.8	12	200
Canada	British Columbia	Riondel (Z)		1.8	8	181
England	Gloucestershire	Cirencester		1.2	15	60
Wales	Cardiganshire	Aberystwyth		1.1	15	600
England	Shropshire	Church Stretton		.5	7	180
England	Cornwall	Feock		.5	4	150
England	Somerset	Glastonbury		.4	5	120
Scotland	Dumfrieshire	Lynhurst		.2	2	140
			Range	.2–23	2–24	
			Mean	1	11	
		C. Potato (Tuber Peeled)				
England	Yorkshire	South Milford		7.6	160	88
Canada	British Columbia	Riondel (S)		2.2	47	500
Canada	British Columbia	Fraser Valley		2.0	33	<0.8
Canada	Newfoundland	Boat Harbour		1.7	38	8
England	Derbyshire	Steetley		.9	29	20
England	Warwickshire	Coleshill		.5	12	110
England	Shropshire	Church Stretton		.4	8	180
England	Middlesex	Hounslow		.4	8	32
England	Sussex	Petworth		.2	5	100
England	Isle of Man	Douglas		.2	4	400
			Range	.2–7.6	4–160	
			Mean	1.6	34	
		D. Carrot (Edible Portion)				
Canada	British Columbia	Riondel (S)		11.0	140	500
Canada	British Columbia	Peachland		9.2	140	23
England	Devonshire	Holsworthy		8.0	120	320
Canada	British Columbia	Nelson		3.2	38	30
England	Cambridgeshire	Cambridge		2.0	17	300
England	Cornwall	Feock		1.2	21	150
England	Gloucestershire	Cirencester		.7	10	60
Scotland	Dumfrieshire	Lynhurst		.6	7	140
Scotland	Selkirkshire	Galashiels		.5	6	150
Scotland	Berwickshire	Coldstream		.2	5	40
			Range	.2–11	5–140	
			Mean	4	50	
		E. Bean (Edible Portion—Various Varieties)				
Canada	British Columbia	Peachland		12	210	23
England	Devonshire	Milton Abbot		6	70	*
Canada	British Columbia	Nelson		5.6	100	30
England	Devonshire	Morwellham		5.0	150	*
Canada	British Columbia	Point Grèy (1967)		4.0	84	2
Canada	British Columbia	Riondel (S)		2.4	25	500
Canada	British Columbia	Point Grey (1968)		2.4	32	1.5
England	Northumberland	Hexham		2.0	40	23
Canada	British Columbia	Riondel (Z)		1.5	25	181
England	Somersetshire	Axbridge		1.0	8	71
			Range	1–12	8–210	
			Mean	4	74	

*not available

TABLE 4. THE MOLYBDENUM CONTENT OF VARIOUS VEGETABLES AND THEIR SUPPORTING SOILS (in ppm)

A. Lettuce (Edible Inner Leaves)

Country	County or Province	Parish or Locality	Dry Plant	Ash	Soil
Canada	British Columbia	Nelson	12	44	<0.8
England	Devonshire	Morwellham	2.0	18	0.8
Canada	British Columbia	Riondel (S)	2	9	2
England	Middlesex	Feltham	.9	5	<0.8
England	Devonshire	Milton Abbot	.8	7	1
Canada	British Columbia	Riondel (Z)	.6	4	<2
England	Kent	Ashford	.4	5	2
Canada	British Columbia	Point Grey (1967)	.3	3	<0.8
Canada	British Columbia	Fraser Valley	.2	1	<0.8
Canada	British Columbia	Point Grey (1968)	<.2	<3	<0.8
		Range	<.2–12	<3–44	
		Mean	<2	<10	

B. Cabbage (Edible Inner Leaves)

Country	County or Province	Parish or Locality	Dry Plant	Ash	Soil
Scotland	Dumfrieshire	Lynhurst	8.0	80	2.0
England	Sussex	Petworth	7.6	90	3.2
Canada	British Columbia	Riondel (Z)	6.4	29	<2.0
England	Somersetshire	Glastonbury	5.2	60	4.4
England	Gloucestershire	Cirencester	4.0	49	2.0
England	Shropshire	Church Stretton	3.8	57	3.6
England	Cornwall	Probus	3.1	33	5.6
Scotland	Stirlingshire	Falkirk	1.6	19	2.0
Scotland	Berwickshire	Coldstream	1.4	20	0.8
England	Kent	Ashford	1.3	13	3.2
		Range	1.3–8	13–90	
		Mean	4	45	

C. Potato (Tuber Peeled)

Country	County or Province	Parish or Locality	Dry Plant	Ash	Soil
England	Warwickshire	Coleshill	2.8	69	2.0
England	Berkshire	Bedford	1.3	43	3.6
England	Somerset	Glastonbury	1.2	30	4.4
Canada	British Columbia	Riondel (S)	.6	14	2.0
Canada	British Columbia	Riondel (Z)	.4	11	<2.0
England	Nottinghamshire	Kirkby-in-Ashfield	.6	10	3.0
England	Derbyshire	Steetley	.3	8	.3
England	Middlesex	Hounslow	.3	5	<0.8
Scotland	Stirlingshire	Falkirk	.2	7	2.0
Wales	Glamorgan	Swansea	.2	7	2.8
		Range	.2–2.8	7–69	
		Mean	.8	20	

D. Carrot (Edible Portion)

Country	County or Province	Parish or Locality	Dry Plant	Ash	Soil
Canada	British Columbia	Riondel (S)	4.2	56	2.0
Canada	British Columbia	Nelson	2.0	24	<0.8
Canada	British Columbia	Peachland	1.8	29	2.5
England	Cambridgeshire	Cambridge	.8	7	2.4
Canada	British Columbia	Riondel (1965)	.8	9	*
England	Devonshire	Holsworthy	<.4	<4	3.2
Scotland	Dumfrieshire	Lynhurst	<.2	<2	2.0
Scotland	Selkirkshire	Galashiels	<.2	<2	1.6
Canada	British Columbia	Fraser Valley (1968)	<.2	<3	?
Canada	British Columbia	Fraser Valley (1966)	+	<2	?
		Range Not Detected	4.2	<2–56	
		Mean	<1	<14	

TABLE 4. (CONTINUED)

E. Bean (Edible Portions—Various Varieties)						
Canada	British Columbia	Peachland		32	550	2.5
Canada	British Columbia	Nelson		28	500	<0.8
Canada	British Columbia	Riondel (S)		18	190	2.0
England	Somerset	Axbridge		9	75	<1.2
Canada	British Columbia	Point Grey (1968)		8.4	110	.8
Canada	British Columbia	Riondel (Z)		5.6	92	<2.0
England	Northumberland	Hexham		3	75	.4
England	Devonshire	Morwellham		2	70	*
England	Devonshire	Milton Abbot		1	10	*
Canada	British Columbia	Point Grey		.9	21	<0.8
			Range	.9–32	10–550	
			Mean	11	169	

*not available
+not detected

TABLE 5. SUMMARY OF DATA FROM TABLES 1 THROUGH 4. THE MEAN AND RANGE OF COPPER, ZINC, LEAD, AND MOLYBDENUM CONTENTS IN SOME REPRESENTATIVE SAMPLES OF LETTUCE, CABBAGE, POTATOES, CARROTS, AND BEANS (in ppm dry material)

	Copper		Zinc		Lead		Molybdenum	
	Mean	Range	Mean	Range	Mean	Range	Mean	Range
Lettuce	9	3–15	86	30–250	12	.3–56	<2	<.2–12
Cabbage	4	1–6	39	10–84	1	.2–2.3	4	1.3–8.0
Potato	3	.5–6	18	8–29	1.6	.2–7.6	.8	.2–2.8
Carrot	3	.6–5	29	12–59	4	.2–11	1	* 4.2
Bean	5.5	3–9	46	30–94	4	1–12	11	.9–32

*not detected

The samples were chosen to represent only the range of elements in oven-dried material, with geographic distribution as wide as was practical with the available data. The general lack of correlation between the amount of an element in a vegetable and the amount in its supporting soil is a matter of some interest to agriculturists and nutritionists alike.

Our methods of analysis, which have been described elsewhere (Warren and Delavault, 1956), make us confident that, although it is possible that some of our results on vegetal matter may be low, it is certain that they are not too high. The analytical data are given as furnished by the laboratory; at best, only two significant figures are reliable.

We have avoided reporting any results from areas known by us to have unusually high morbidity or mortality patterns. It is fair to state, however, that in virtually every area where unusual epidemiological features have been reported to us, usually in connection with cancer or multiple sclerosis, or both, we have found some striking deviations from what we consider to be the "normal" trace-element relationships in one or more vegetables growing in that area. There may be a relative excess of zinc or molybdenum in some of the vegetables growing in that area; in other cases there may be much more lead than copper.

DISCUSSION OF RESULTS

All the above data should be considered only as a reconnaissance and much more data must be assembled before any conclusions should be attempted. We have already explored the possibility of using other soil extractants, such as aqua regia, and acetic acid at *p*H 4.7; we have compared surface soils and deep soils; we have screened soils, and we have used "raw" soils; and we have determined *p*H and organic content of soils. All of these factors have their influence on relationships between the trace elements occurring in vegetal matter and its supporting soil.

Having stated the above, and without support by statistical analyses, we think that the following hypotheses may act as guides for some future investigations:

(1) There is no simple relationship between the trace-element content of a soil and the vegetables growing on that soil.

(2) Each vegetable tends to have a characteristic mineral uptake which usually compensates to a considerable extent for an excess or deficiency of a particular element in a soil.

The manner in which bean and cabbage plants can extract molybdenum from molybdenum-poor soils is quite striking. Conversely, vegetables growing on soils with abnormally high lead contents do not always pick up as much lead as might be anticipated.

(3) If one excludes the heavily mineralized samples from Riondel, it seems clear that British soils have a much higher lead content than normal Canadian soils. What proportion of this lead is the result of pollution, we are unable to estimate. However, evidence already presented by Patterson (1965) and Warren (1965) suggests that is is an important fact.

(4) If one accepts the general proposition that some elements can inhibit the uptake of another element, for example, molybdenum is known to upset the

TABLE 6. THE COPPER, ZINC, LEAD, AND MOLYBDENUM CONTENT OF LETTUCE SAMPLES FROM THE PARISH OF BERE FERRERS
(in ppm oven dried material)

Sample No.	Year	Copper	Zinc	Lead	Molybdenum
"A"	1962	8	170	40	2.0
2 (a)	1963	8	66	1	<0.6
2 (b)	1963	4	94	2	<0.6
2 (d)	1963	16	330	250	3.0
2 (e)	1963	13	120	23	5.0
2 (f)	1963	7	71	7	0.6
2 (g)	1963	10	170	10	0.6
2 (h)	1963	15	260	32	0.6
#5	1964	50	180	50	1.5
#4	1964	10	120	44	2.0

normal uptake by sheep of copper in some herbage, then abnormal amounts of such elements as lead, mercury, or molybdenum in vegetal matter may possibly affect the uptake by man of some other element, such as copper.

(5) Normally, whether one considers the Earth's crust, soils, fish livers, or average man, there tends to be a crude, but distinct, relationship between the concentrations of a few of the trace elements, and particularly those with which we have been dealing in this paper. Normally there is more zinc than copper in the vegetables we have examined, more copper than lead, and more lead than molybdenum.

Our limited experience prevents us arriving at any conclusions, but the few samples that follow should serve to indicate why we suggest that nutritionists should carefully investigate any food that has trace-element contents which deviate markedly from the normal.

SOME EXAMPLES OF TRACE-ELEMENT IMBALANCES IN LOCALITIES WHERE ANOMALOUS EPIDEMIOLOGICAL FEATURES HAVE BEEN REPORTED

The College of General Practitioners Records and Statistics Unit (1966) has reported that the morbidity rate for malignant neoplasms is significantly higher in the Tamar Valley than in the Stoke-on-Trent area, an area considered to be relatively free from concentrations of heavy metals. The late Allen-Price (1960) drew attention to the above-normal cancer mortality rates in some particular parishes in West Devon including Bere Ferrers, which is in the Tamar Valley. Dr. Allen-Price sent us lettuce samples in three succeeding years. These were analyzed in our laboratories, and results are listed in Table 6.

It will be noted that, if one excludes the Riondel samples, the lead and zinc results are well above any normal collection and are in sharp contrast to those found in the next set of samples from parishes with a below-normal mortality rate from malignant neoplasms (Table 7). Incidentally, Allen-Price's data covered a 20-year period.

Although the above data deal with lettuce, similar results can be and have been observed in other vegetables.

TABLE 7. THE COPPER, ZINC, LEAD, AND MOLYBDENUM CONTENT OF LETTUCE SAMPLES FROM THE PARISHES OF BRADSTONE, CORYTON, LYDFORD, MILTON ABBOT, AND WALKHAMPTON
(in ppm oven dried material)

Sample No.	Year	Copper	Zinc	Lead	Molybdenum
3 (a)	1963	10	69	5	0.6
6 (a)	1963	15	78	10	3.0
13 (a)	1963	5	66	7	6.0
17 (a)	1963	10	87	10	<0.6
#6	1964	13	120	9	0.8
—	1962	3	50	6	2.0
25 (a)	1963	6	61	20	5.0
(b)	1963	5	61	3	6.0

In conclusion it should be mentioned that the trace-element content of vegetal matter varies greatly with the season in which it is harvested. Mitchell and Reith (1966) have cited variations of the order of twenty to one in the lead content of gramens. In an effort to minimize this particular problem, all the vegetable samples were collected at approximately the same stage of development.

CONCLUSIONS

The trace-element content of mature vegetables varies by factors ranging from three to one to as much as thirty to one. Using such epidemiological data as are readily available, it is interesting to note that, in localities where unusual disease patterns have been described in medical literature, high concentrations of trace elements, often associated with imbalances, have been noted in garden soils and in the food crops grown on them.

ACKNOWLEDGMENTS

The authors must first and foremost acknowledge their deep indebtedness to the Donner Canadian Foundation for financial support which made it possible to collect so many samples from such widely separated areas and make so many analyses. In addition, we were enabled to carry out tests on artificially and industrially "salted" soils. Our colleagues in Agriculture, notably Professors C. A. Rowles and L. E. Lowe in Soil Science and V. C. Brink in Plant Science, not only gave much valuable advice, but also placed some of their facilities at our disposal. Dr. W. K. Fletcher also made some constructive suggestions.

Many people in both British Columbia and Great Britain helped us to make our necessary collections. To these people and to the many members of the Royal College of General Practitioners of Great Britain who participated, we gratefully acknowledge our indebtedness.

The background data on which this work was based resulted from funds provided by the National Research Council of Canada (A 1805), the Geological Survey of Canada, and Kennco Explorations (Western) Limited, supplemented by most welcome contributions from four other mining companies.

REFERENCES CITED

Allen-Price, E. D., 1960, Uneven distribution of cancer in West Devon: Lancet, v. 1, p. 1235–1238.

College of General Practitioners Records and Statistics Unit, 1966, Some contrasts in morbidity distribution: College of General Practitioners Jour., v. 1, p. 74–83.

Mitchell, R. L., and Reith, J. W. S., 1966, The lead content of pasture herbage: Jour. Sci. Food Agriculture, v. 17, p. 437–440.

Patterson, Clair C., 1965, Contaminated and natural lead environments of Man: Archives of Environmental Health, v. 11, p. 344–360.

Warren, Harry V., 1965, Medical Geology and Geography: Science, v. 148 (3669), p. 534–539.

Warren, Harry V., and Delavault, R. E., 1956, Soils in geochemical prospecting: Western Miner and Oil Review, v. 29 (12), p. 36–42.

MANUSCRIPT RECEIVED BY THE SOCIETY DECEMBER 5, 1969

Printed in U.S.A.

THE GEOLOGICAL SOCIETY OF AMERICA, INC.
MEMOIR 123, 1971

Trace Elements Related to Specific Chronic Diseases: Cancer

ARTHUR FURST
Institute of Chemical Biology, University of San Francisco
San Francisco, California

ABSTRACT

All living tissues are composed mainly of eleven elements, but to remain viable, minute amounts of a few elements of the transition series also must be present. These act as mediators of the biocatalysts, the enzymes. The trace elements that have been most extensively studied are: Fe, Cu, Mn, Mg, Mo, and Zn. The body as it ages concentrates a large number of other elements; many of these, when present in excess, have been reported as being responsible for the induction of cancer, but adequate documentation exists for only a small number of metals. Unequivocal experiments reveal that *nickel, cadmium,* and some *chromium* compounds are true metal carcinogens. Suggestive, but inadequate, evidence exists for *cobalt* and *lead. Selenium* compounds are questionably carcinogenic according to our present data. *Arsenic* has been strongly indicted as a primary human carcinogen. *Asbestos* may prove to be a carrier for the carcinogenic metals, nickel and chromium. No tumors have ever been induced in experimental animals with either *iron* or *copper.*

The mechanisms of metal carcinogenesis are not known. Theories for this action must account for how metals may get into cells—probably by first combining with a small protein and, as this neutral complex, entering the cell by a process known as *pinocytosis.* Once in the cell, the metal complex can dissociate and the free ion can combine with sulfur-containing units in enzymes. This new metal derivative can completely inhibit the enzyme action or modify the kinetics of the enzyme. Under the influence of these conditions, cells may grow at different rates.

The carcinogenic metals also may combine with nucleic acids and change the genetic information transferred to new cells. "Metal-toxicity" can be a general manifestation of these phenomena, or, in selected cases, the deranged metabolism may give rise to "malignant" new growth, which represents cancer.

CONTENTS

INTRODUCTION

Living tissue, be it animal or vegetable, is composed mainly of eleven elements; these can be divided into two groups: five metals and six nonmetals. These elements, the so-called "bulk" elements are: H, C, N, O, Na, Mg, P, S, Cl, K, Ca. For those species requiring hemoglobin, iron can be added to this list.

In addition to the "bulk" elements, there are several other elements, mainly from the transition series in the periodic chart, which living cells require in order to function and remain viable. Present only in minute quantities, and thus often called *trace elements,* these metals help to regulate the dynamics of the life processes. The ones most extensively studied for their effects on biological systems are: Cu, Zn, Mn, Co and Mo. These five elements, along with the macro-nutrient elements Mg, Ca, and Fe, function alone or in some combination with each other as mediators of many biocatalysts, the enzymes (Bowen, 1966).

The role of these mediator elements is varied. Some elements may function to activate an enzyme and help bind the substrate to the enzyme. Here, the ion is not an unique obligate; in the case of a few enzymes, one metal may be replaced by another, providing the replacement is in the same oxidation state and is present at approximately the same concentration. The replacement of magnesium by manganese ions in some phosphate transferases or decarboxylases is an example of the interchangeability of ions in some systems.

In other enzyme systems, the ions are bonded chemically to some specific portion of the enzyme as a complex or chelate, in a definite stoichiometric ratio. The metal is considered an integral part of this vital biocatalyst. Zinc- and copper-containing enzymes are known as *metalloenzymes;* if a metal such as iron or cobalt is bound by a porphyrin group, the enzyme is called a *metalloporphyrin enzyme.* Another group of enzymes contains the following metals alone or in pairs, Fe, Cu, Mn, Mg, Mo, or Zn. These latter enzymes function as dehydrogenases and may be classified as *metalloflavin enzymes.* In these systems the metals are complexed with varying degrees of affinity. Some are removed easily by EDTA solutions; some require more drastic treatment. In many cases the chemical means required to remove the metal may destroy the enzyme.

In spite of their importance—all life depends on them—only minute numbers of atoms of the essential trace elements are required in living tissues.

For example, only one copper atom may be present in some tissues per 10^7 atoms of all other atoms. These ratios may vary from tissue to tissue. A reasonable estimate of the ratio of Cu:Mo:Co present in the human body is in the order of 1000:100:1.

Accumulation of Metals by Tissues

The human body is capable of concentrating not only the essential elements, but also other elements. Studies on autopsy specimens from nondiseased human beings revealed that various organs contained at least twenty other elements in addition to the expected trace elements. This accumulation of metals may be part of the phenomenon of aging, for many of these metals are absent in tissues of stillborns. Few, if any, of these metals have significant biological or physiological action except, perhaps, for toxic effects in relation to inhibition of certain enzymes. The accumulative metals include: Be, Al, Si, Ti, V, Cr, Ni, As, Se, Sr, Pd, Ag, Cd, Sn, Ba, and Au (Tipton and Shafer, 1964). Although these elements are distributed widely in the body, some metals apparently have a proclivity for concentrating in specific tissues (Molokhia and Smith, 1967; Morgan and others, 1960). Much of this information has been obtained by use of the emission spectrograph. Now, with the introduction of more refined analytical techniques, namely, the atomic absorption spectrometer and the electron probe microanalyzer, more accurate values may be obtained on specific tissues. Studies currently underway may require modification of this list of accumulated metals, and it is possible that metal shifts from one body organ or tissue to another, within the same animal, will be found.

One interesting aspect of metallic accumulation is that the amount of metal per gram of tissue varies widely from one geographic region to another. For example, cadmium and lead values are related to the section of the country where the samples were taken. All blood samples tested from residents of

Vermont had cadmium; not so in New York (Kubota and others, 1968). Man is exposed to many of these chemicals as a result of his civilized environment. Other elements may come to us from the soil by way of the plant foodstuffs we eat. A summary of previous work in this area is given by Bowen (1966).

Despite the fact that most of the metals found in tissues has no clear biological function at the present time, some metals, for example, Cr, Ni, and Se, are always found in tissues and, in some cases, in a definite molar ratio to each other. As more is learned in the field of nutrition, some of these metals, which now are considered nonessential, may be found necessary for specific cellular functions, albeit in extraordinarily minute quantities. Two of these metals, not now classified as necessary for any enzyme function, can modify the kinetics of a few enzyme reactions. Examples are given in Table 1.

TABLE 1. EFFECT OF NICKEL AND CHROMIUM IONS ON ENZYME SYSTEMS

Activity	Enzymes affected
	Nickel
Functions in the enzyme but is easily replaced	Alkaline phosphomonoesterase I Acid phosphomonoesterase II
Facultative activator	Acid phosphomonoesterase III
Inhibitor of enzyme	Fructo-1,6 -diphosphatase Enolase
	Chromium
Functions in the enzyme but is easily replaced	Arginase-a Urease
Facultative activator	Arginase-b Carboxylase Pyruvate decarboxylase
Inhibitor of enzyme	Histidase

Metal Carcinogenesis

Even if most of these accumulated metals are nonessential, their importance cannot be minimized, for there is evidence that the presence of an excess amount of a metal in an organ may result in the induction of a malignant tumor. Thus, in addition to a variety of organic compounds, which include some polycyclic hydrocarbons, azo dyes, aromatic amines, and many nitrosamines (Clayson, 1962), a few free metals and some of their compounds can induce cancer in both rodents and human beings (Hueper and Conway, 1964).

The field of metal carcinogenesis is complicated; many claims are made about a variety of active elements, but adequate documentation exists for only a few. The list of metals which have been reported as potential carcinogens includes: Al, Cr, Fe, Co, Ni, Cu, Zn, As, Se, Ag, Cd, Sn, Hg, and Pb (Furst, 1963). In a recent publication, critically reviewing the field, a suggestion is made that many of these agents may not be true metal carcinogens, but may act nonspecifically as solid irritants (Furst, 1967).

Because this subject has not been extensively studied (though it is not really new), and because the experimental procedures used in some investigations left much to be desired, it is not surprising that some textbook writers have questioned the validity of the view that metals can act as true carcinogens (Clayson, 1962). Hueper is one worker who has consistently listed metals in his compilations of carcinogens (Hueper and Conway, 1964) and, where known, has designated the exact chemical nature of the active metal or its compound.

Perhaps one of the deterrents in accepting the fact that some metals can induce malignancies is the difficulty that arises in trying to distinguish between nonspecific solid-state carcinogenesis (Bischoff and Bryson, 1964) and true metal carcinogenesis. Various histological types of tumors have been produced in rodents after some plastic sheets have been placed under the skin (Oppenheimer and others, 1952). Also, many solids implanted into muscles can cause sarcomas, possibly as a result of local hypoxia or anoxia, or by the contact of tissues with smooth surfaces (Oppenheimer and others, 1956). Some years ago it was shown that normal myocardial cells maintained in tissue culture and kept under hypoxic conditions transformed to malignant cells (Goldblatt and Cameron, 1953).

Recent work, however, points to the conclusion that some metals and their compounds are true carcinogens. Pure metallic *nickel* and many of its compounds can be considered as causes of cancer (Payne, 1964). In Figure 1, the formulae of some nickel compounds found to be carcinogenic are presented. Included in this list are soluble, insoluble, and moderately soluble compounds. Three of these compounds were investigated in our laboratories (Haro and others, 1968). The target organ for this element is usually the muscle.

Another well-documented carcinogenic element is *cadmium*, either as the metallic powder (Heath and others, 1962), the oxide, the sulfide (Kazantzis and Hanbury, 1966), or even in a soluble form (Gunn and others, 1963; Roe, 1964). In some cases the malignant tumors appear in a tissue distant from the application site, and, as in the case of nickel, the cadmium-induced tumors are transplantable (Heath and Webb, 1967). Some ores of both of these metals are also active carcinogens (Hueper and Conway, 1964).

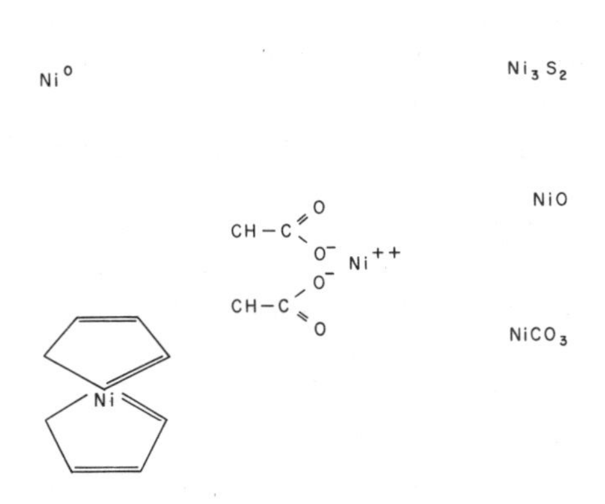

Figure 1. Formulae of nickel compounds found to be carcinogenic.

Iron powder has not been found to be carcinogenic in our laboratory

(Haro and others, 1968). In the form of an organic dextran complex, however, it has induced local tumors at the site of injection in rabbits after repeated injections (Haddow and others, 1964). Hematite has been implicated as a carcinogen for many miners who have developed lung cancer (Faulds and Stewart, 1956). Large amounts of pure *cobalt* metal, when implanted into rodent muscle tissue, also produce local sarcoma at the injection site (Heath, 1960).

Chromite ore roasts and a few selected chromium compounds have induced malignant tumors in subcutaneous and muscle tissues as well as in the lungs of both mice and rats. The nature of the malignant tumor produced was dependent on the chromium compounds used and the site of implantation (Grogan, 1957, 1958; Hueper and Conway, 1964; Payne, 1960). Not enough work has been done on the pure chromium metal or on some of the chromium compounds now in industrial use to draw firm conclusions.

Many more experimental studies must be carried out before we can include or exclude some metals as true carcinogens. Clues may come from epidemiologists, but to establish unequivocally the activity of any pure metal, animal experiments must be performed. For example, although there is no substantial evidence that *copper* can cause cancer, a high incidence of lung cancer has been noted among coppersmiths (Agnese and others, 1959). In our laboratories copper powder failed to induce any type of tumor, thus, at the present time, this metal should not be classified with the carcinogens.

Many *arsenic* compounds have been implicated as being carcinogenic to human beings. Workers exposed to sprays and dusts containing arsenic have a higher cancer incidence rate than the population as a whole. This is especially true of those who work in the vineyards (Roth, 1956). A paradox exists here for, to date, no biological model exists; this can only be attributed to the fact that the human being is uniquely sensitive to this element. With every other known carcinogen, it has been possible to induce some type of malignancy in one experimental animal or more (Clayson, 1966). Previous attempts to produce arsenical cancer in animals have been reviewed (Furst and Haro, 1969); this publication lists a set of criteria which should be considered before a metal is considered a true carcinogen. Within the last decade, new experiments that attempted to induce cancer in rodents with inorganic arsenicals also failed (Baroni and others, 1963). Hueper and Payne (1962), who exposed rodents to arsenic trioxide in drinking water over a period of 24 months, could not come to a conclusion as to the carcinogenicity of this arsenic compound in these rodents.

In spite of the lack of biological model, Hueper and Conway (1964) present a number of arguments to support the view that inorganic arsenic is a primary cause of cancer in man. Frost (1967), on the other hand, after working with arsenicals as additives for feed, concludes that the toxicity of arsenic in some studies may be confused with that of the selenium, and that the evidence against arsenic as a carcinogen is by no means conclusive. Attempts should be continued in searching for a biological model, and studies should be initiated to evaluate, in rodents, both organic and inorganic arsenic compounds of varying oxidation states. Different routes of administration should be tried.

Selenium presents a complicated picture. Some workers found that feeding seleniferous grain to rodents produced cancer of the liver (Nelson and others, 1943). However, in more recent experiments in which selenium was administered in the diet as either sodium selenite (Se^{+4}) or sodium selenate (Se^{+6}), no tumors were produced (Harr and others, 1967). It is obvious from these conflicting findings that more work must be done with selenium in its various forms before any firm decision can be reached as to its carcinogenicity.

Many scientists doing research in metal carcinogenesis may not have appreciated the fact that a variety of solids, regardless of their chemical composition, may cause local cancer when imbedded subcutaneously, and that metal foils with smooth surfaces are especially active (O'Gara and Brown, 1967). Thus the *silver, per se,* of silver foil may not be a true carcinogen, because its action is related to a "smooth surface" effect (Oppenheimer and others, 1956). *Lead* should be reinvestigated, for until now very few substantiated papers have shown that lead compounds could induce tumors. Until the early 1960s most lead studies gave negative results in regard to cancer (Schroeder and others, 1964). In the past ten years, however, several papers have reported that lead acetate, whether injected in the rodent or administered in the diet, will induce tumors, especially in the kidney (Boyland and others, 1962; Esch and others, 1962).

There is, at the present time, a resurgence of interest in the carcinogenic potential of *asbestos* (Hinson, 1965; Williams, 1965). Both asbestos miners and men working with insulation materials have an unusually high incidence of a type of cancer called *mesothelioma,* which often arises in the pleura, but also develops from peritoreum or pericardium. For the asbestos-cancer problem, in contrast to the smoking-cancer problem, a rodent model does exist (Lynch and others, 1957); tumors can be induced in the lungs of fowl following the injection of asbestos particles (Peacock and Peacock, 1965). Asbestos, although containing substantial amounts of oil which can be tumor-promoting (Roe and others, 1966), also contains both nickel and chromium in addition to iron. Evidence now is accumulating that implies that asbestos may serve only as a carrier of the carcinogenic metals (Dixon and others, 1969). Also, there are different minerals in the general class of asbestos; studies are now being conducted to find which variety is most potent as a carcinogen.

An unexpected finding in our laboratory is that *titanium* metal, as the pure powder, will induce fibrosarcomas in an inbred strain of rats and lymphosarcomas in others (both tumors have not occurred in the same animal). To date none of our control animals has had a spontaneous lymphoma. The element titanium is not usually associated with biological materials nor does it have a known natural biological function, although this element is concentrated in certain human organs. Using the electron probe microanalyzer, high levels of titanium were recently found in leukocytes derived from bone marrow; and especially in cases where lymphoblastic lymphoma involved the marrow (Carroll and Tullis, 1968).

Much more experimental work is required to establish, without doubt, the carcinogenicity of other metals listed in the introduction. The pure metal as well as its inorganic and organic compounds should be tested. Different species of animals, as well as different routes of administration, should be employed. It is best to avoid the subcutaneous route of implanting solids, because the "growths" induced may be difficult to classify and at all times the possibility of nonspecific solid-state carcinogenesis must be kept in mind.

The Imbalance of Metals

Human beings exposed to environmental pollution may accumulate abnormal amounts of metals in various body organs. Excess accumulation of metals that are not usually associated with biological function may lead to toxic symptoms, or even to death; drinking water containing some unusual minerals may also contribute to higher cancer rates (Stocks and Davies, 1960). One possible consequence of abnormal exposure to metals is trace-metal imbalance, a condition that may be fostered by the ingestion of foods grown in soils deficient in essential elements or in soils with an unusually high content of one metal.

A complex interrelationship has been shown to exist between soil, plants, and animals (Underwood, 1962). Tomatoes grown in a mineral-deficient medium develop tumors (Swain and Reir, 1968). The omission of barium from a refined diet given to rats will result in a growth depression, while trace amounts of certain elements in the food stock (like arsenic in pig feed) may give rise to larger and fatter animals. In natural surroundings, animals feeding upon minerally imbalanced plants may develop pathological conditions. Plants grown in soil deficient in cobalt have been implicated for their neurotoxic action in sheep. In selenium-rich areas, this element concentrates in plants, and cattle eating these plants become afflicted with sclerosis.

To trace these relationships in humans, it is best to study a population whose members have resided for many years in one region. It is hard to find such populations in the United States, as this country is a nation of nomads. Also, the people of the United States come from many racial stocks. A cancer survey of 185 million Americans, such as that reported by Shimkin (1968a), does not shed much light on the soil-cancer problem. Information from other countries, however, does imply that cancer rates may be related to differences in the metal content of the soil.

Warren, a geologist, has suggested that geology may hold a key to better health. He gives details on the cancer rate of three regions in the British Isles and suggests that areas with Devonian rocks have a higher incidence of the disease than those underlain by rocks of other ages (Warren, 1962). Stomach cancers have been found in greater proportions in populations that live in regions where the Zn/Co ratio and the Zn/Cu ratio exceeded definite values (Stocks and Davies, 1960). A recent study in South Africa of the Transkei Country gave evidence that the unusually high esophageal-cancer incidence in the Bantu is associated with mineral deficiency in soils and, hence, garden plants (Marais and Drewes, 1962). Also, a deficiency of molybdenum and an excess of several other ions were noted (Burrell and others, 1966). In a 1964 paper, Stocks and Davies suggest that an important factor leading to metal imbalance in human beings may be the presence of a metal that antagonizes the utilization of either zinc or copper. An excess of such an antagonizer in the soil may have the same end result as a soil deficiency of one of these essential metals. All of these observations suggest that this is a study area of much potential in the environmental health field. Cannon (1970) emphasizes the need for an "interdisciplinary team approach" to study the relationship of trace metals and their availability to plants, animals, and man. Such cooperative studies might shed new light on the contribution of natural and polluted environments to health and disease.

EXPERIMENTS

Materials and Methods

In all experiments described here, commercially purchased fine metal powders of at least 200 mesh were used. Powders of 99.9-percent purity were suspended in an oil vehicle, a synthetic triglyceride, tricaprylin (also called trioctanoin), which was redistilled under reduced pressure before it was used. The collected fraction boiled over a 2°C range.

Animals were weanling inbred Fischer-344 rats. For each test an equal number of males and females, generally 25 each, was used. There were two sets of controls, each of an equal number of rats. The pure vehicle was injected in one set; the second set was left alone to become the "shelf controls."

Procedures

Prior to the initiation of the long-term carcinogenicity test, an acute (10 day) and chronic (10 week) toxicity study was carried out for each metal or compound evaluated. The criteria for toxicity were: weight loss, mortality, and gross and microscopic pathological changes observed during necropsies. These toxicity tests served to determine the maximum dose of the metal which the animals could tolerate during the period of the carcinogenic studies.

For the carcinogenesis experiments, all rats were injected in the thigh muscle of the right hind leg with 0.2 ml of the oil vehicle suspension containing the fine metal powder. Vehicle-control animals received 0.2 ml of trioctanoin without the metal. "Shelf" controls were not injected. All animals received water and chow as they desired. Aseptic techniques were used; sterile disposable needles and syringes were used to administer the vehicle and suspensions. When the rats were small, they were housed five per cage; when they grew larger, they were kept in individual cages. Treatments were continued monthly until nodules appeared at the injection sites, or unless infection or inflammation was noted.

During the course of the experiment, all animals were carefully examined two or three times a week at the time they were put into clean cages with fresh shavings. Each rat was weighed weekly for the first three months and then monthly until the experiment was terminated. When definite nodules appeared in more than one animal, the monthly injection of the metal suspension was terminated, but the examination, weighing, and cage changing continued until the animals became moribund. Autopsies were conducted on every animal and suspicious tissues were sent to a consulting pathologist for histological diagnosis. Each animal was examined for metastasis. In all cases in which tumors were induced, attempts were made to transplant some neoplastic tissue into new host animals of the same strain.

The experiments were terminated when approximately five animals were left alive in each group, with the exception of the controls which were sacrificed after 900 days.

Results and Discussion

The data from a series of representative experiments are given in Table 2. Listed are the pure metals tested, the day when the last animals were sacrificed in each experiment, the dose used of each metal for both females and males (note dose in mg per animal), and the type and number of malignancies induced in each sex. Both nickel[1] and titanium metals gave positive results. The latent period before the malignant tumors appeared varied from 4 to 6 months for nickel to about 16 months for titanium. Copper and iron failed to induce tumors. No spontaneous sarcomas or lymphomas have appeared in any control animal in a three-year period. The Fischer-344 inbred rat does, however, have a high spontaneous incidence of ovarian or testicular tumors.

Plates and photomicrographs are representative of the results found. Figure 2 shows a rat bearing a transplanted malignant ascites lymphoma which was induced by titanocene. Figure 3 is a photomicrograph of the spleen of the rat shown in the previous photograph. Figure 4 presents a rat with the transplanted ascites, and a rat with a nickel-induced fibrosarcoma at the site of injection of the metallic agent. A photomicrograph of this spindle-cell fibrosarcoma is shown in Figure 5.

POSSIBLE MECHANISM OF ACTION

Cell Penetration

The actual mechanism of action of metals as carcinogens remains unknown to date. It is assumed that before a metal can act on, or modify, any natural substances like enzymes or nucleic acids, the metal must penetrate the cell-

[1]In other experiments, not described here, a number of nickel compounds were tested for their carcinogenic properties. In each case over 50 percent of the animals developed fibrosarcomas at the site of application. Every tumor-bearing rat had metastatic lesions in the lungs. Tumors were selected at random and were transplanted into new host animals; no tumor failed to grow on transplantation. Similar results were obtained with titanocene, an organo-titanium compound; however, the titanium-induced tumors were more difficult to transplant.

Table 2. Cancers Induced by Selective Metals

Compound*	Total days on test	Dose†		Fibrosarcomas				Lymphomas			
		M	F‡	M	F	Total	%	M	F	Total	%
Trioctanoin (control)	955	C	C	0	0	0	0	0	0	0	0
Trioctanoin	917	C	C	0	0	0	0	0	0	0	0
Nickel	520	31	19	18	20	38	76	0	0	0	0
Titanium	820	39	23	2	0	2	4	3	0	3	6
Copper	470	45	45	0	0	0	0	0	0	0	0
Iron	573	75	75	0	0	0	0	0	0	0	0
Chromium	644	100	100	1	0	1	2	1	0	1	2
Aluminum	516	75	75	(no tumors, still on test)							

*All metals tested were powders suspended in oil (trioctanoin).
†Doses are given in mg/rat.
‡In each experiment 25 male and 25 female rats were used.

Figure 2. Left: Normal control rat. Right: Fischer-344 rat bearing 7-day transplant of malignant ascites lymphoma induced by treatment with titanocene dichloride.

wall barrier. The metal in all probability, does not enter the cell as the pure metal (oxidation state of zero) but as a complex of the ion of that metal, because positively charged ions do not move freely across cell membranes. The cell membranes are composed of a complex pattern of proteins, lipids, and polysaccharides. Permeability of ions is selective, and an "electrotype-pump mechanism" has been hypothesized to explain the fact that certain univalent ions, such as sodium and lithium, enter cells with difficulty, although potassium and cesium can penetrate these cells with relative ease. Biological

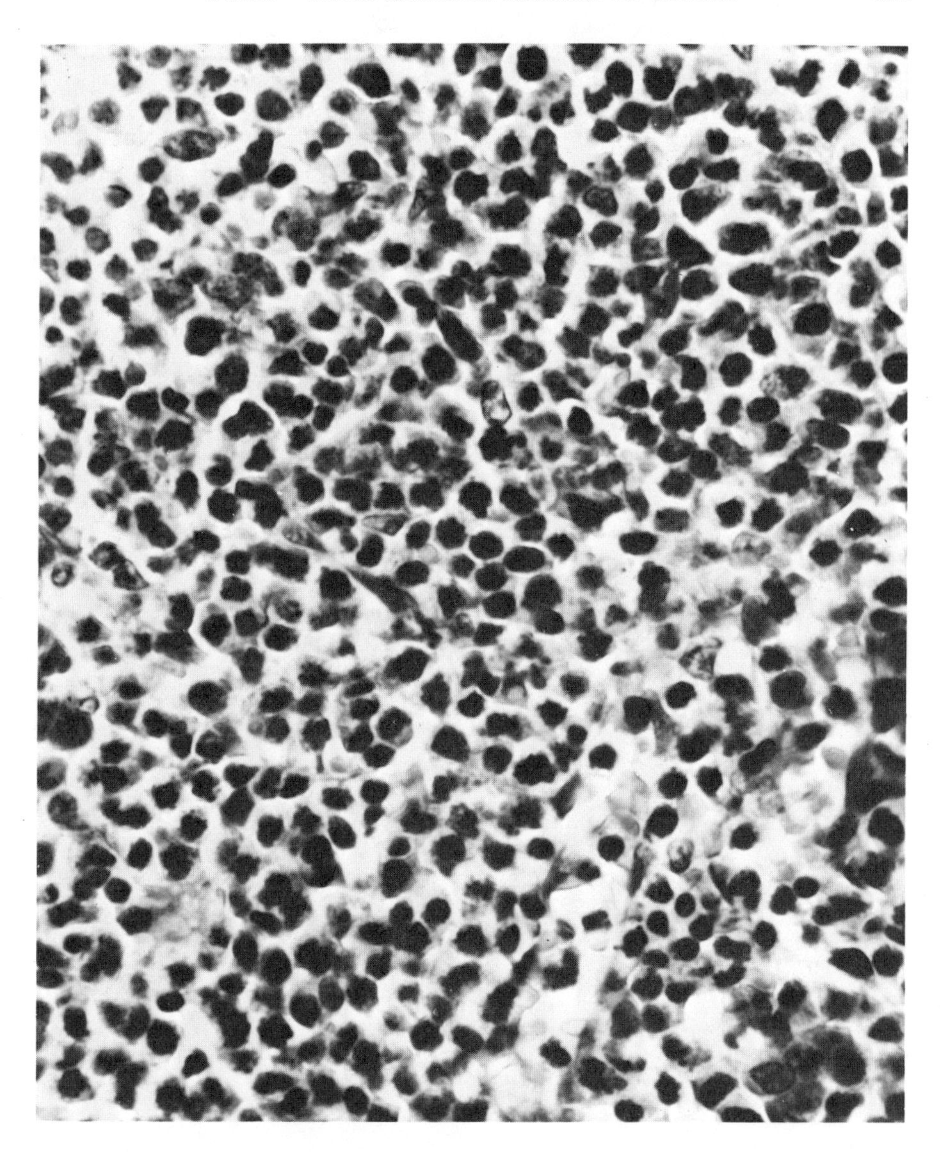

Figure 3. Photomicrograph of spleen, showing titanocene dichloride-induced malignant lymphoma. Note infiltration by mononuclear cells (×400).

work must be done in order to keep the concentration of these ions different within and without the cells, and this energy is derived from the metabolic activities of the cell. The difference in concentration of the ions around the inner and outer surface of the cells sets up an electrical potential. This can act as a barrier to prevent the free movement of those ions of the transition series into the cells of tissues.

To account for the fact that these ions are actually found in cells, it is most reasonable to assume the formation of an electrically neutral complex with

Figure 4. Left: Fischer-344 rat bearing 7-day transplant of malignant ascites lymphoma induced by treatment with titanocene dichloride. Right: Fischer-344 rat bearing large nickelocene-induced spindle cell sarcoma at site of intramuscular injection in right hind leg. Note asymmetry from tumor weight.

some biochemical moiety that can effectively bind the metallic ion and form a non-ionic compound. The carcinogenic metals have varying affinities for proteins, and also different rates of entry into cells (Hueper and Conway, 1964). Small proteins can be the complexing agent, for there are many active sites on various proteins which can chelate the metal ion. Once formed, if the compound (a metal proteinate) is of the right molecular weight and dimensions, it can be engulfed by cells by a phenomenon known as *pinocytosis*.

The hypothesis that we are developing presupposes that metals, to be effective carcinogens, must first be dissolved in the tissues. Evidence from our laboratories has shown that nickel does dissolve in the muscle. It is not possible to predict from chemical solubility data alone the degree of solution of a

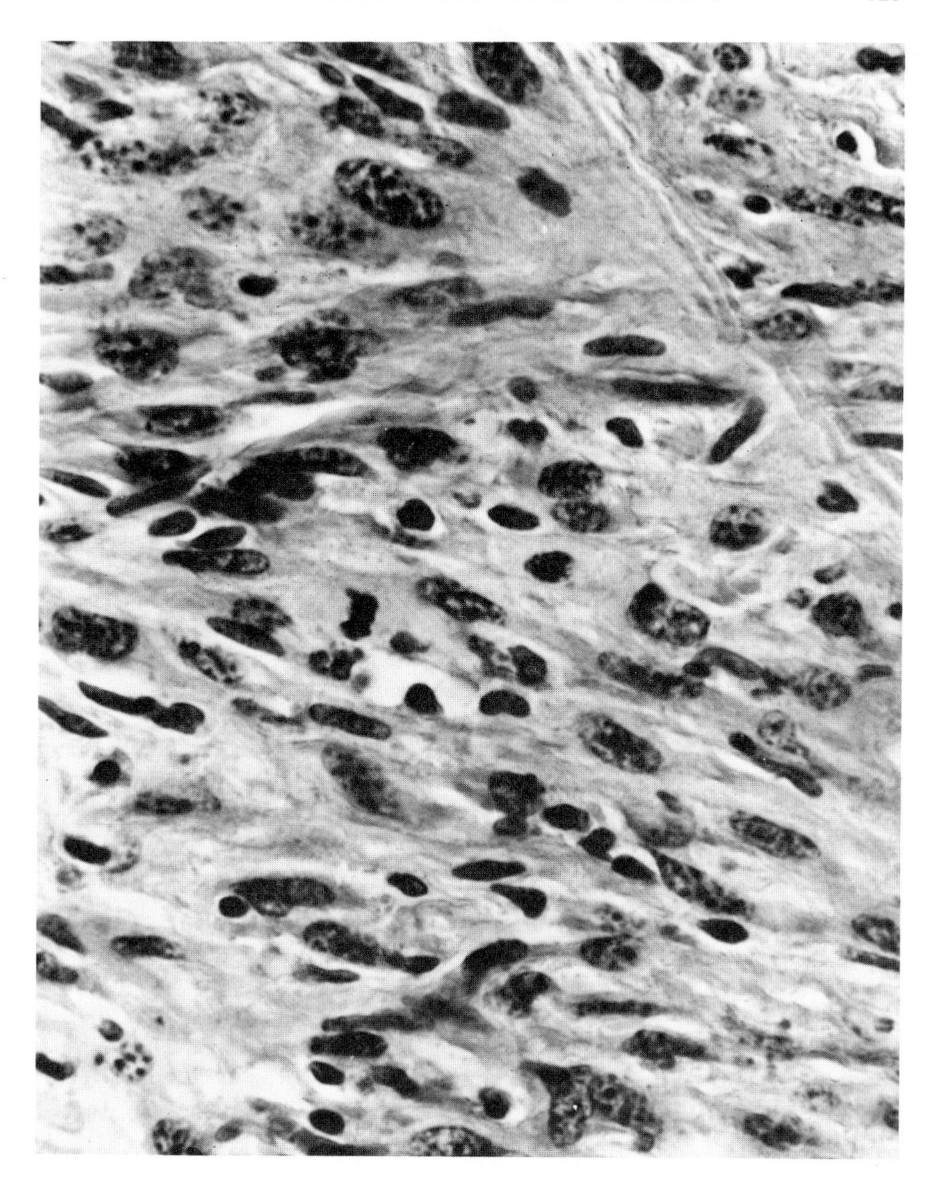

Figure 5. Photomicrograph of nickelocene-induced spindle-cell sarcoma removed from site of intramuscular injection in right hind leg of Fischer-344 rat (×400).

metal in this tissue. Nickel powder (200 mesh) was suspended in trioctanoin and injected into the muscle of a group of rats. The animals were placed in individual metabolism cages, and urines were collected. The control rats received only the vehicle trioctanoin. Figure 6 represents the daily excretion patterns of nickel ions for three nickel-injected and three control rats. The concentration of nickel in the urine of treated animals was much higher than predicted.

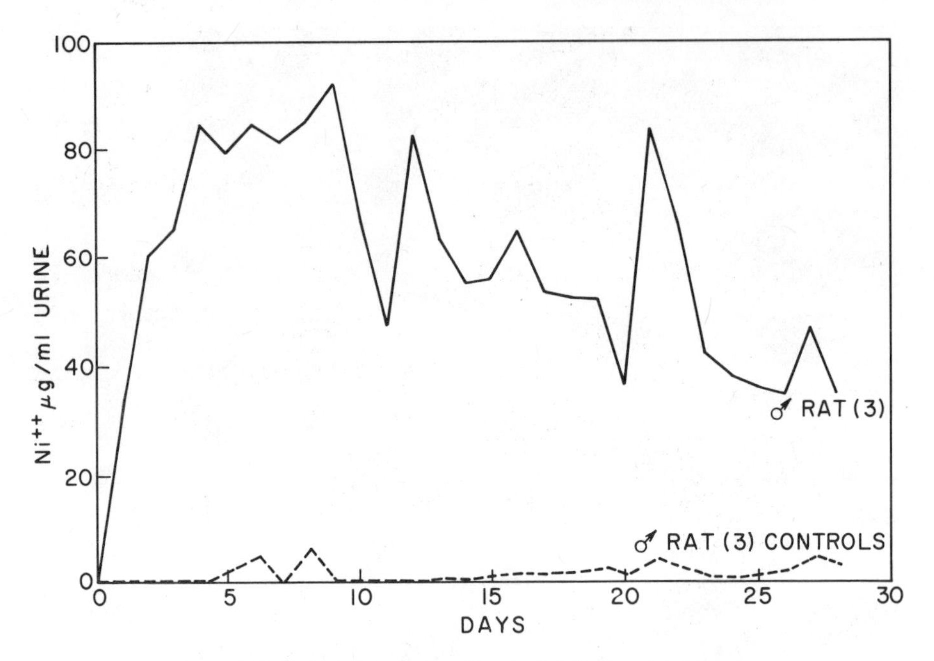

Figure 6. Daily nickel excretion after single injection nickel powder.

Apparently all metals can be dissolved by tissues. Even titanium, which shows very low corrosion properties in chlorine solutions, dissolved when implanted into the rabbit muscle (Ferguson and others, 1960).

Action on Enzymes

The metal complexes may dissociate after penetrating the cell wall, leaving a free ion that can react in a number of ways. Ions like chromium, in a high oxidation state, may be reduced and then enter into and modify oxidative mechanisms (Hueper and Conway, 1964), or they may replace (partially) a normal trace element, resulting in a metal imbalance. Heavy metal ions may combine with active centers in cell constituents, such as amino acids or sulfide groups.

It is well known that many metals form sulfide compounds easily; these compounds are among the most stable and least soluble of all chemical combinations. These same ions can easily complex with sulfhydryl groups (–SH) of organic compounds. These sulfur groups, also known as thiols, are found in proteins, some amino acids, small peptides, and over 50 percent of the enzymes.

The biochemical groups which contain one thiol group usually form a redox system of the nature: –SH + –S–S–. Free –SH groups are needed for normal cell division; the formation of a metal combination (–S–M–S–) thus can interfere with this process.

As the metallic ions under consideration are polyvalent, they should not have much effect on enzymes which contain only one sulfhydryl group. However, profound changes in enzyme activity can be expected if the enzyme in question has two thiol groups arranged geometrically so that the metallic ion can combine with both to form an intra –S–M–S– bond. Biochemical transformations dependent upon these enzymes would be impeded. Either the expected products would not be formed if the enzyme is completely inhibited, or the rate of formation of the product would be modified. The end result of the combination of metal ions with sulfhydryl groups in these enzymes would be to retard product formation, or to modify the energy relationships, or to change the rate of protein or nucleic-acid syntheses. The most clear-cut manifestation of the combination of metal ions with sulfur-containing enzymes is "metal poisoning." This manifestation would apply to all metals with an affinity for sulfur compounds; additional biological properties would be expected of metals that cause cancer, which may be a modification of enzyme reaction rates with different results on various cells. Under these altered growth patterns, neoplastic cells may emerge.

Suggestions for Nucleic Acid Interactions

Carcinogenic metals may have a more fundamental biological property. The ion, after migrating from the complexing agent which carried it into the cell, may bind to some nucleic acid. All nucleic acids contain minute quantities of some metals (Fuwa and others, 1960; Wacker and others, 1959), the function of which may be related to the conformation of these nucleic acids. "Abnormal" metals, or abnormal amounts of "normal" metals, may bind with nucleic acids and change their size, degree of polymerization (Butzow and Eichhorn, 1965), or the genetic message transmitted to new cells.

Various functions of chelate compounds on biological systems have been reviewed (Eichhorn, 1961; Haddow, 1959). If some metal, for example, iron, is essential to stabilize nucleic acid during the process of cell division (Ivanov, 1965), it is conceivable that the metals that tend to selectively accumulate in some tissues (Molokhia and Smith, 1967) and may displace many of the iron ions and bind preferentially to nucleic acids in large enough quantities (Sunderman and others, 1962) to distort the original nucleic acid-iron complex. Nickel ion, which has a great tendency to form ammonia-type complexes, can easily combine with the base, guanidine, contained in all nucleic acids. During cell division, the rate or nature of the nucleic acid replication may be altered. As nickel would not be expected to dissociate readily after the cell has divided, the nucleic acid strands could not be stabilized by hydrogen-bond formation in the same way as before. This might result in a new nucleic-acid sequence. A variant in replication could then take place which could be manifest as a genetic change, resulting in the formation of a cancer cell. Once such genetic changes occurred, the presence of nickel ions would no longer be required.

Alternate Hypotheses

Not all metals can induce cancer, and not all essential metals can be replaced by a foreign ion. In the case of manganese metabolism, copper or zinc could displace the essential metal in some systems, but not in others (Cotzias, 1960).

Our hypothesis also predicts that the electronic configuration of the metal ion is *not* a factor in its carcinogenicity. For example, among the carcinogenic-nickel compounds listed in Figure 1 are a variety of oxidation numbers and electronic configurations. Therefore, neither a study of coordination chemistry (Basolo and Johnson, 1964), nor atomic radii (Hueper and Conway, 1964) seems to be particularly helpful when attempts are made to interpret the mechanism of nickel carcinogenesis.

Other modes for carcinogenic action can be ascribed to some metals. Irritants, like dusts, can enhance activity of (other) known carcinogens. Mice continuously exposed to Al_2O_3 dust (1–5 μg particle size) developed more lung adenomas after administration of 4-nitroquinoline-N-oxide than an equal number of controls given the carcinogen, but not exposed to the dust (Kohayashi and others, 1968).

Some metals may act indirectly. For example, human beings are constantly being exposed to carcinogenic hydrocarbons from our polluted atmosphere. The lungs and liver inactivate these hydrocarbons by an induced enzyme system called benzpyrene hydroxylase. Nickel inhibits this enzyme (Sunderman, 1967); thus, it may permit the hydrocarbon to exist for longer periods allowing a longer time for the carcinogen to interact with the tissues. Another hypothesis (suggested by others) calls attention to the fact that carcinogenic hydrocarbons can form undefined complexes with certain cellular constituents such as riboflavin; these combinations, called *charge transfer complexes* (Szent-Gyorgyi and others, 1960), permit the transfer of electrons to acceptors—a condition that may be necessary in the causation of cancer. Metals, too, have been found to form charge transfer complexes (Szent-Gyorgyi, 1961).

Perhaps one of the most interesting suggestions to explain an indirect mechanism for carcinogenic action and the relation of human cancer to mineral-deficient soils, as found in the Transkei area of South Africa, is as follows: In order for nitrates to be reduced to amines for plant utilization, the essential element Mo must be present. In Mo-deficient soil, only partial reduction of the nitrates may occur, and *nitrosamines* rather than amines may be the end product (Burrell and others, 1966). Nitrosamines are among the most potent organic carcinogens known, producing cancer in specific organs of the body at very low doses.

These are the various hypotheses which have been set forth to help explain metal carcinogenesis. None has been substantiated to date; however, they are all valuable in that they point to logical directions for further investigation.

ACKNOWLEDGMENTS

I should like to acknowledge Mr. Richard T. Haro and Mr. Michael Schlauder, Senior Biologists on this program. Their knowledge, help, and devotion are much beyond thanks. Mr. David Cassetta has worked as our assistant on all phases of the project since his high school days. Also my thanks to Mr. Tommie Granger who has kept superb animal quarters.

The experimental work was done under contract with the National Cancer Institute, no. Ph-43-64-886. Project Officers were Drs. C. Baker, and U. Saffiotti, and now Dr. Melvin Rueber.

The diMartini Fund for Cancer Research at the University of San Francisco permitted us to make studies incidental to, but not part of, the National Cancer Institute contract.

REFERENCES CITED

Agnese, T., Veris, B. De, and Santolini, B., 1959, Incidence of lung cancer in relation to occupation in Genoa: Igiene Mod., v. 52, p. 149–160.

Baroni, C., Van Esch, G. J., and Saffiotti, U., 1963, Carcinogenesis tests of two inorganic arsenicals: Am. Medical Assoc. Arch. Environmental Health, v. 7, p. 668–674.

Basolo, F., and Johnson, R. C., 1964, Coordination Chemistry, *in* The Chemistry of Metal Complexes: New York, W. A. Benjamin, p. 22–59.

Bischoff, F., and Bryson, G., 1964, Carcinogenesis through solid state surfaces: Experimental Tumor Research Prog., v. 5, Basel/New York, Karger, p. 85–133.

Bowen, H. J. M., 1966, Trace Elements in Biochemistry: London, Academic Press, 241 p.

Boyland, E., Dukes, C. E., Grover, P. L., and Mitchley, B. C. V., 1962, The induction of renal tumours by feeding lead acetate to rats: British Jour. Cancer, v. 16, p. 283–288.

Burrell, R. J. W., Roach, W. A., and Shadwell, A., 1966, Esophageal cancer in the Bantu of the Transkei associated with mineral deficiency in garden plants: Jour. Natl. Cancer Inst., v. 36, p. 201–209.

Butzow, J., and Eichhorn, L., 1965, Interactions of metal ions with polynucleotides and related compounds. IV. Degradation of polyribonucleotides by zinc and other divalent metal ions: Biopolymers, v. 3, p. 95–107.

Cannon, H. L., 1970, Trace element excesses and deficiencies in some geochemical provinces in the United States, *in* Hemphill, D. D., *Editor,* Trace Substances in Environmental Health-III: Columbia, Missouri, Univ. Missouri, p. 21–43.

Carroll, K. G., and Tullis, J. L., 1968, Observations on the presence of titanium and zinc in human leucocytes: Nature, v. 217, p. 1172–1173.

Clayson, D. B., 1962, Chemical Carcinogenesis: Boston, Little, Brown and Co., p. 118.

——, 1966, The induction of cancer by chemicals, *in* Ambrose, E. J. and Roe, F. J. C., *Editors,* The Biology of Cancer: London, D. Van Nostrand Company, Ltd., p. 157–175.

Cotzias, G. C., 1960, Metabolic relations of manganese to other minerals: Federal Proc., v. 19, p. 665–668.

Dixon, J. R., Lowe, D. B., Richards, D. E. and Stockinger, H. E., 1969, The role of trace metals in chemical carcinogenesis-asbestos cancers, *in* Hemphill, D. D., *Editor,* Trace Substances in Environmental Health-II: Columbia, Missouri, Univ. Missouri, p. 141–159.

Eichhorn, G. L., 1961, Metal chelate compounds in biological systems: Federal Proc., v. 20, p. 40–51.

Esch, G. J. Van, Genderen, E. van, and Vink, H. H., 1962, The induction of renal tumors by feeding of basic lead acetate to rats: British Jour. Cancer, v. 16, p. 289–297.

Faulds, J. S., and Stewart, M. J., 1956, Carcinoma of the lung in haematite miners: Jour. Path. Bacteria, v. 72, p. 353–366.

Ferguson, A. B., Laing, P. G., and Hodge, E. S., 1960, The ionization of metal implants in living tissue: Jour. Bone and Joint Surgery, v. 42, p. 77–90.

Frost, D. V., 1967, Arsenicals in biology—retrospect and prospect: Federal Proc., v. 26, p. 194–208.

Furst, A., 1963, The Chemistry of Chelation in Cancer: Springfield, Illinois, Charles C Thomas, p. 15–19.

Furst, A. and Haro, R. T., 1969, A survey of metal carcinogenesis: Experimental Tumor Research Prog., v. 11, Basel/New York, Karger, p. 102–133.

Fuwa, K., Wacker, W. E. C., Druyan, R., Bartholomay, A. F., and Vallee, B. L., 1960, Nucleic acids and metals, II: Transition metals as determinants of the conformation of ribonucleic acids: Natl. Acad. Sci. Proc., v. 46, p. 1298–1307.

Goldblatt, H., and Cameron, G., 1953, Induced malignancy in cells from rat myocardium subjected to intermittent anaerobiosis during long propagation *in vitro*: Jour. Experimental Medicine, v. 97, p. 525–552.

Grogan, C. H., 1957, Experimental studies in metal carcinogenesis. VIII. On the etiological factor in chromate cancer: Cancer, v. 10, p. 625–638.

——, 1958, Experimental studies in metal carcinogenesis. XI. On the penetration of chromium into the cell nucleus: Cancer, v. 11, p. 1195–1203.

Gunn, A., Gould, T. C., and Anderson, W. A. D., 1963, Cadmium-induced interstitial cell tumors in rats and mice and their prevention by zinc: Jour. Natl. Cancer Inst., v. 31, p. 745.

Haddow, A., 1959, The possible role of metals and of metal chelation in the carcinogenic process, *in* Wolstenholme, O. E. W., and O'Connor, M., *Editors*, CIBA Foundation Symposium on Carcinogenesis, Mechanism of Action: Boston, Little, Brown and Co., p. 300–307.

Haddow, A., Roe, F. J. C. and Mitchley, B. C. V., 1964, Induction of sarcomata in rabbits by intramuscular injection of iron-dextran ("Inferon"): Brit. Medical Jour., v. 1, p. 1593–1594.

Haro, R. T., Furst, A., Payne, W. W., and Falk, H., 1968, A new nickel carcinogen: Am. Assoc. Cancer Research Proc., v. 9, p. 28.

Harr, J. R., Bone, J. F., Tinsley, I. J., Weswig, P. H , and Yamamoto, R. S., 1967, Selenium toxicity in rats. II. Histopathology, *in* Selenium in Biomedicine: Westport, Connecticut, Avi, p. 153–178.

Heath, J. C., 1960, The histogenesis of malignant tumours induced by cobalt in the rat: British Jour. Cancer, v. 14, p. 478–482.

Heath, J. C., Daniel, M. R., Dingle, J. T., and Webb, M., 1962, Cadmium as a carcinogen: Nature, v. 193, p. 592–593.

Heath, J. C., and Webb, M., 1967, Content and intracellular distribution of the inducing metal in primary rhabdomyosarcoma induced in the rat by cobalt, nickel and cadmium: British Jour. Cancer, v. 21, p. 768–779.

Hinson, K. F. W., 1965, Cancer of the lungs and other diseases after exposure to asbestos dust: British Jour. Chest Diseases, v. 59, p. 121–128.

Hueper, W. C., and Conway, W. D., 1964, Chemical Carcinogenesis and Cancers: Springfield, Illinois, Charles C Thomas, p. 379–402.

Hueper, W. C., and Payne, W. W., 1962, Experimental studies in metal carcinogenesis. Chromium, nickel, iron, arsenic: Am. Medical Assoc. Arch. Environmental Health, v. 5, p. 445–462.

Ivanov, V. I., 1965, Role of metals in deoxyribonucleic acid: Biofizika, v. 10, p. 11–16.

Kazantzis, G., and Hanbury, W. J., 1966, Induction of sarcoma in the rat by cadmium sulfide and by cadmium oxide: British Jour. Cancer, v. 20, p. 190–199.

Kohayashi, N., Ide, G., Katsuki, H., and Yamare, Y., 1968, Effect of aluminum compounds on the development of experimental lung tumor in mice: Gann, v. 59, p. 433–436.

Kubota, J., Lazar, V. A., and Losee, F., 1968, Copper, zinc, cadmium and lead in human blood from 19 locations in the United States: Am. Medical Assoc. Arch. Environmental Health, v. 16, p. 788–793.

Lynch, K. M., McIver, F. A., and Cain, J. R., 1957, Pulmonary tumours in mice exposed to asbestos dust: Arch. Industrial Health, v. 15, p. 207–214.

Marais, J. A. H., and Drewes, F. F. R., 1962, Relationship between solid geology and oesophageal cancer distributed in the Transkei: South Africa Geol. Survey Ann. Rept., v. 1, p. 105–114.

Molokhia, M. M., and Smith, H., 1967, Trace elements in the lung: Am. Medical Assoc. Arch. Environmental Health, v. 15, p. 745–750.

Morgan, K. Z., Snyder, W. S., and Auxier, J. H., 1960, Report of I.C.R.P. Committee II on permissible dose for internal radiation (1959): Health Physics, v. 3, p. 148–150.

Nelson, A. A., Fitzhugh, O. G., and Calvery, H. O., 1943, Liver tumors following cirrhosis caused by selenium in rats: Cancer Research, v. 3, p. 230–236.

O'Gara, R. W., and Brown, J. M., 1967, Comparison of carcinogenic actions of subcutaneous implants of iron and aluminum in rodents: Jour. Natl. Cancer Inst., v. 38, p. 947–957.

Oppenheimer, B. S., Oppenheimer, E. T., Danishefsky, I., and Stout, A. P., 1956, Carcinogenic effect of metals in rodents: Cancer Research, v. 16, p. 439–441.

Oppenheimer, B. S., Oppenheimer, E. T., and Stout, A. P., 1952, Sarcomas induced in rodents by imbedding various plastic films: Soc. Experimental Biology Proc., v. 79, p. 366–369.

Payne, W. W., 1960, The role of roasted chromite ore in the production of cancer: Am. Medical Assoc. Arch. Environmental Health, v. 1, p. 20–26.

——, 1964, Carcinogenicity of nickel compounds in experimental animals: Am. Assoc. Cancer Research Proc., v. 5, p. 50.

Peacock, P. R., and Peacock, A., 1965, Asbestos-induced tumors in white leghorn fowls: New York Acad. Sci. Ann., v. 132, p. 501–503.

Roe, F. J. C., 1964, Cadmium neoplasia: Testicular atrophy and Leydig cell hyperplasia and neoplasia in rats and mice following the subcutaneous injection of cadmium salts: Brit. Jour. Cancer, v. 18, p. 674–681.

Roe, F. J. C., Walters, M. A., and Harington, J. S., 1966, Tumour initiation by natural and contaminating asbestos oils: Intnat. Jour. Cancer, v. 1, p. 491–495.

Roth, F., 1956, Uber die Chronische Arsenvergifung der Moselwinzer uber besonderer Berucksichtigung des Arsenkrebses: Zeit. fur Krebs., v. 61, p. 287–319.

Schroeder, H. A., Balassa, J. J., and Vinton, W. H., Jr., 1964, Chromium, lead, cadmium, nickel, and titanium in mice: Effect on mortality, tumours and tissue levels: Jour. Nutrition, v. 83, p. 239–250.

Shimkin, M. B., 1968a, Distribution of cancer in the United States: Am. Medical Assoc. Arch. Environmental Health, v. 16, p. 503–512.

Shimkin, M. B., 1968b, Environmental carcinogenesis: Am. Medical Assoc. Arch. Environmental Health, v. 16, p. 513–530.

Stocks, P., and Davies, R. I., 1960, Epidemiological evidence from chemical and spectrographic analysis that soil is concerned in the causation of cancer: Brit. Jour. Cancer, v. 14, p. 8–22.

——, 1964, Zinc and copper content of soils associated with the incidence of cancer of the stomach and other organs: Brit. Jour. Cancer, v. 18, p. 14–24.

Sunderman, F. W., Jr., 1967, Inhibition of benzpyrene hydroxylase by nickel carbonyl: Cancer Research, v. 27, p. 950–955.

Sunderman, F. W., Jr., Horn, R. S., and Sunderman, F. W., 1962, Binding of nickel and other trace metals by ribonucleic acid: Federal Proc., v. 21, p. 311.

Swain, L. W., and Reir, J. P., Jr., 1968, Tumorigenesis on mineral-deficient tomato plants: Cancer Research, v. 28, p. 2496–2501.

Szent-Gyorgyi, A., 1961, Metals and charge transfer: Natl. Acad. Sci. Proc., v. 47, p. 1398–1399.

Szent-Gyorgyi, A., Isenberg, I., and Baird, S. L., Jr., 1960, The electron donating properties of carcinogens: Natl. Acad. Sci. Proc., v. 46, p. 1444–1449.

Tipton, I. H., and Shafer, J. J., 1964, Statistical analysis of lung trace element levels: Am. Medical Assoc. Arch. Environmental Health, v. 8, p. 58–67.

Underwood, E. J., 1962, Trace Elements in Human and Animal Nutrition: New York, Academic Press, p. 363–383.

Wacker, W. E. C., Warren, E. C., and Vallee, B. L., 1959, Chromium, manganese, nickel, iron and other metals in ribonucleic acid from diverse biological sources: Jour. Biol. Chemistry, v. 234, p. 3257–3262.

Warren, H. V., 1962, Does geology hold a key to better health?: Mining Engineering, v. 14, p. 41–45.

Williams, W. J., 1965, Asbestosis and lung cancer: Am. Medical Assoc. Arch. Environmental Health, v. 10, p. 44–45.

MANUSCRIPT RECEIVED BY THE SOCIETY MAY 18, 1970

Printed in U.S.A.

THE GEOLOGICAL SOCIETY OF AMERICA, INC.
MEMOIR 123, 1971

Geographic Patterns in the Risk of Dying

HERBERT I. SAUER
AND
FRANK R. BRAND, M.D.
Ecology Field Station, Heart Disease and Stroke Control Program, Public Health Service, Columbia, Missouri

ABSTRACT

There were marked differences among geographic areas within the United States, as well as among countries, in the risk of dying in middle age, during the period 1959 to 1961. The lowest rate areas are principally in the Great Plains; some are in the mountains. The highest rate areas, mostly near the East Coast, have all-causes death rates approximately double those of the low-rate areas. These patterns are similar to those for 1949 to 1951. Several foreign countries have rates for middle-aged males definitely lower than the rates for any of the 509 state economic areas in the United States. (All rates presented are race-specific, for whites.)

The geographic patterns of female rates tend to be similar to those for males, but with the lowest rate areas concentrated in Oklahoma and Texas and very high rates for some Middle Atlantic areas. The patterns for *malignant neoplasms* are similar, but not identical, to those for the *cardiovascular-renal diseases* and *coronary heart disease. Rheumatic heart disease, chronic nonspecific respiratory diseases,* and *accidental and violent deaths* have geographic patterns independent of each other and of any other cause category analyzed.

Many factors have been suspected of being responsible for these geographic differences in the risk of dying, but, with the possible exception of cigarette smoking, available evidence for these various factors is at present inconclusive.

The potential significance of these differences is substantial. More specifically, if the rates of the lowest rate areas had applied to the entire United States in 1959 to 1961, there would have been approximately 100,000 fewer deaths per year under age 65 alone.

CONTENTS

INTRODUCTION

Substantial geographic differences exist in the risk of dying in middle age, for all causes as well as for various specific causes. Our objective is to present some of these patterns and to consider necessary procedures for testing hypotheses regarding factors responsible for these differences. This work has been built upon the foundation laid by Enterline and Stewart (1956; Stewart and Enterline, 1957) a dozen years ago as well as by the National Center for Health Statistics (1949–1966) and various state health departments.

The risk of dying is lowest in childhood and increases drastically with age. Rates for natural causes—that is, for all causes except accidents, homicides, and suicides—show an increase of roughly ten percent with the passing of each year of age from early adult life to age 95, thus forming an approximation of a straight line on a semi-logarithmic scale. In studying geographic patterns of disease, it is therefore necessary to control carefully for age through the use of age-specific or age-adjusted rates.

For white females, the percentage increase is relatively constant, whereas for white males the increase is greatest in early middle life. This causes a bulge upward in the death curve for middle age, which may arbitrarily be defined either as age 45 to 64 or as age 35 to 74.

METHODS

Special tabulations were made, in accord with our specifications, by age, sex, and race for 18 cause categories by the National Center for Health Statistics, and rates were calculated for the 3-year period 1959 to 1961 for the 509 state economic areas (SEA's) of the United States, as defined by the U.S. Bureau of the Census (1963, p. ix–x).

There are 206 metropolitan and 303 nonmetropolitan SEA's. The metropolitan SEA's are similar to standard metropolitan statistical areas (SMSA's), differing in that (1) an SEA lies wholly within one state and wholly within one economic subregion; (2) a few of the smaller SMSA's are classified as part of a nonmetropolitan SEA; and (3) in New England, SEA's consist of one or more metropolitan counties. A nonmetropolitan SEA generally consists of 6 to 20 contiguous counties that are relatively homogeneous as to the type of agricultural, industrial, and commercial activities and various demographic factors.

Rates are derived from the average annual number of deaths by place of usual residence divided by the population at risk. Death rates for ages 45 to 64 and 35 to 74 have been age-adjusted by the direct method (Linder and Grove, 1943, p. 66–69), by 10-year age groups, to the total U.S. population in those age groups in 1950. Thus, standard vital-statistics methods were used, with further refinements dictated by the needs for the study of human ecology. For areas with substantial resident institutional populations, adjustments have been made in the estimated population at risk in a manner similar to that used in a prior study (Sauer and others, 1966, p. 456).

Our focus is upon middle age for several reasons: (1) in that age range the causes of death tend to be specific, (2) the rates are high for males as compared with females, and (3) these deaths may be described as premature. Earlier in life, particularly under age 35, the number of deaths is too small for easy study of differences among many areas. Later in life, roughly age 75 and over, the geographic patterns frequently differ from those in middle age (Enterline and Stewart, 1956, p. 851). Data for nonwhites tend to present geographic patterns similar to those for whites, but this group presents particular problems, methods for which need further study (U.S. Bureau of the Census, 1964a, 1964b; Sauer, 1966; National Center for Health Statistics, 1968). For simplicity and specificity of presentation, all rates presented in this paper are for whites only.

Mortality data for the United States seem to be of high quality, even though admittedly not perfect (Sauer and Enterline, 1959, p. 521). Broad cause categories such as the *cardiovascular-renal (CVR) diseases* are more likely to be accurate then lesser causes. Even so, it seems appropriate to check for rarely occurring complementary patterns of mortality; for example, when the CVR death rate is low and the rate for non-CVR causes is high, it is appropriate to explore the possibility that some CVR deaths are in fact being classified as due to *symptoms, senility,* and *ill-defined causes* or other non-CVR categories. The study of patterns of specific, narrowly defined categories of cause of death may offer greater ecological promise, but their use may also increase the risk of differences in vocabulary and classification of cause from one area to another.

The *International Classification of Diseases* (7th revision) by the World Health Organization (1957) is the source of the three-digit codes of causes of death used in this presentation.

ALL CAUSES

Death rates for middle-aged whites vary substantially from state to state (Enterline and Stewart, 1956). For 1959 to 1961 the pattern is similar to that for 1949 to 1951 for white males aged 45 to 64. The states in the quartile with the lowest rates are all west of the Mississippi River except Wisconsin, with rates from 11.6 to 13.4 per 1000 population. The states in the quartile with the highest rates (from 15.4 to 19.2) are mostly along the East Coast, from Massachusetts to Georgia; Nevada, Arizona, and Louisiana are also included. States, however, are often not very satisfactory as units for epidemiological and ecological studies, not only because of their large size, but particularly because of their marked lack of homogeneity in general population characteristics and death rates. For example, the New Orleans and Chicago areas are quite different from those of the remainder of the states in which they are located.

The use of the 509 SEA's for presentation of *all-causes* death rates for middle-aged white males presents a pattern somewhat similar to that by state (Fig. 1) but with differences resulting from the additional detail given by the larger number of areas. The two lowest octiles of rates are comparable with the lowest quartile in other reports. The lowest octile of areas—the 64 areas with the lowest rates—is predominantly in and near the Great Plains, particularly in the West North Central States, while most of the 64 highest rate areas, the highest octile, are near the East Coast, especially in the Southeast. The second-lowest and second-highest octiles follow similar patterns.

Rates for middle-aged white females present patterns that to a considerable extent are similar to those for males but are by no means identical (Fig. 2).

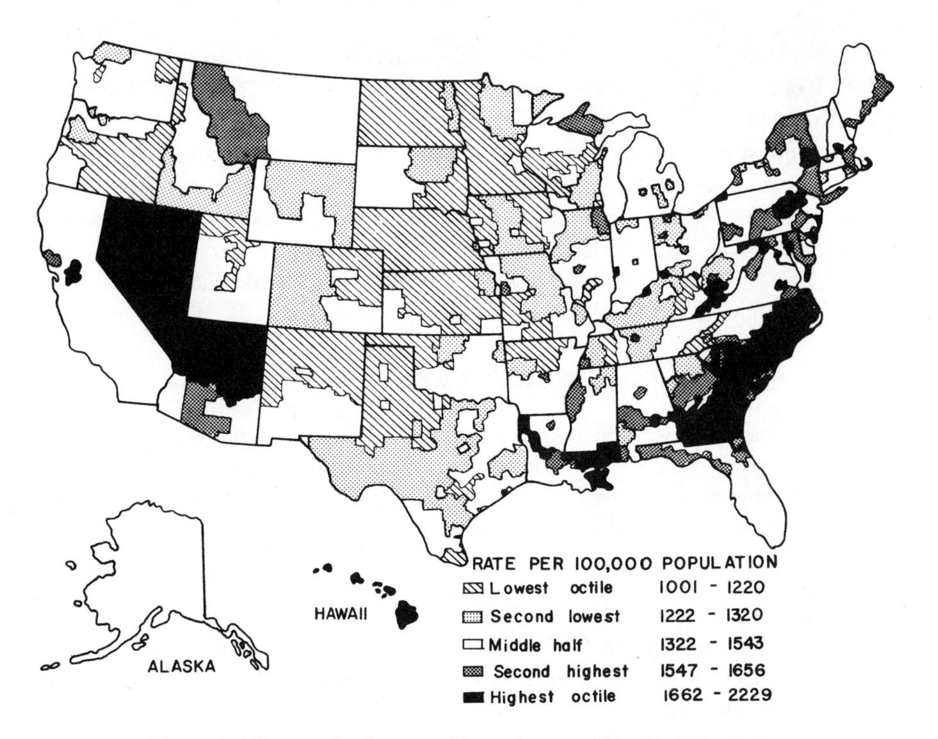

Figure 1. All-causes death rates, white males, age 45 to 64, 1959–1961.

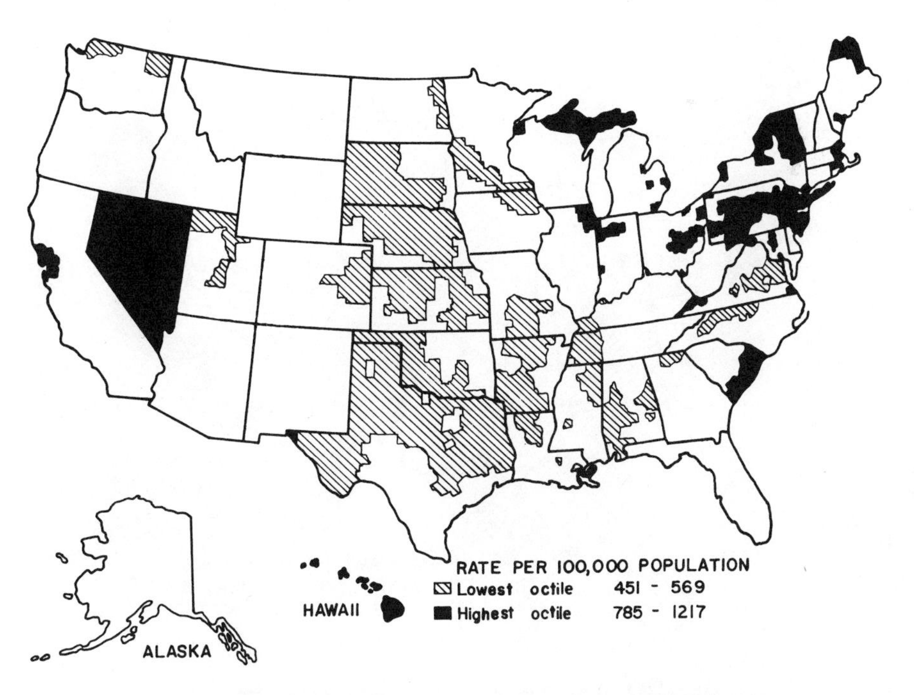

Figure 2. All-causes death rates, white females, age 45 to 64, 1959–1961.

For white females, the lowest rate areas occur more frequently in the West South Central States and the highest rate areas occur more frequently in the Northeast and nearby areas.

The highest rate areas have death rates about twice as high as the lowest rate areas. This generalization applies for females as well as for males. In this age group (45–64) for whites, male rates are on the average more than double the female rates. Thus, in spite of marked geographic differences, only two SEA's, one in New Jersey and one in Hawaii, have female rates higher than the lowest male rate (in Nebraska).

CARDIOVASCULAR-RENAL DISEASES

The diseases of the heart and blood vessels, or the *cardiovascular-renal (CVR) diseases,* include a wide variety of entities, many of which have obscure or unknown causes, for example, *hypertensive disease, atherosclerosis,* and *glomerular nephritis.* This is true for other chronic diseases as well.

Males

Male rates for age 45 to 64 for the CVR diseases present a geographic pattern similar to that for all causes, as would be expected, since more than one-half of all deaths are allocated to this category (Fig. 3). As a group, however, the non-CVR causes show a moderate degree of similarity to the pattern for the CVR diseases.

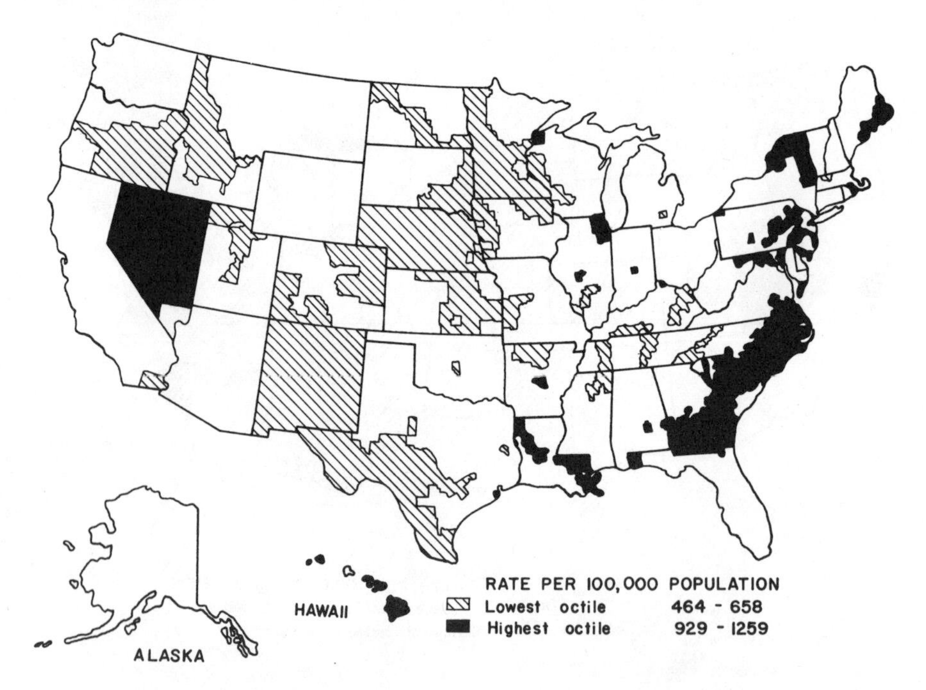

Figure 3. CVR death rates, white males, age 45 to 64, 1959–1961.

Along the East Coast, the highest rate areas are both metropolitan and nonmetropolitan; elsewhere the highest rate areas tend to be metropolitan. As an extreme illustration, the high rate for Nevada is due to the very high rate for the Reno Metropolitan area (Duffy and Carroll, 1967, p. 137). Although there are exceptions, metropolitan areas generally have slightly higher rates than do nonmetropolitan areas (Enterline and others, 1960, p. 760).

The areas with the lowest rates are mostly in the Great Plains and the Rocky Mountains, with a few in the mid-South. Several of the low-rate areas west of the Mississippi River are metropolitan.

The rate distribution patterns for CVR diseases, as well as for all causes, persist over a period of time. That is, the 1959 to 1961 patterns are similar to those for 1949 to 1951 (Enterline and others, 1960) and 1939 to 1941 (Gover, 1949) in spite of differences in analytical methods.

Compared with other countries, rates for middle-aged males in the United States are high for CVR diseases as well as for all causes (Table 1). Three entire countries, the Netherlands, Norway, and Sweden, and the Jewish population of Israel, have lower death rates for all causes than do any of the lowest rate areas in the United States. In order to be more specific, rates in this table are for whites only and include countries with CVR-diseases rates accepted in a report by the American Heart Association (1965, p. 21) plus South African whites. The rates appear to be reasonably comparable in quality. However, for France about 8 percent of the deaths are listed as "cause unknown."

For all causes of death, more than 40 areas in the United States have rates higher than the highest for any of the listed foreign countries; for CVR diseases, more than 100 areas have higher rates than do any of these countries.

Females

The female mortality pattern for cardiovascular diseases is to a substantial extent similar to that for all causes, with the highest rates concentrated in the Middle Atlantic States but with a few high-rate areas in the Carolinas and elsewhere (Fig. 4).

Middle-aged white females in the United States have rates that compare quite favorably with those in other countries (Table 1) in contrast to the high rates for white males.

There are 25 areas in the United States with all-causes white-female rates lower than for any entire foreign country. Most of these areas are in Oklahoma and north-central Texas. Two areas have lower CVR rates. On the other hand, only 9 areas had higher rates for all causes combined and 14 had higher CVR rates than did any foreign country.

Both the metropolitan and nonmetropolitan areas of Hawaii have very high rates for whites (male and female) for all causes as well as for CVR diseases. The possibilities for selective migration should not be ignored, nor should the relatively small white population—suggesting the desirability of suspending judgment as to the possible influence of the Hawaii environment.

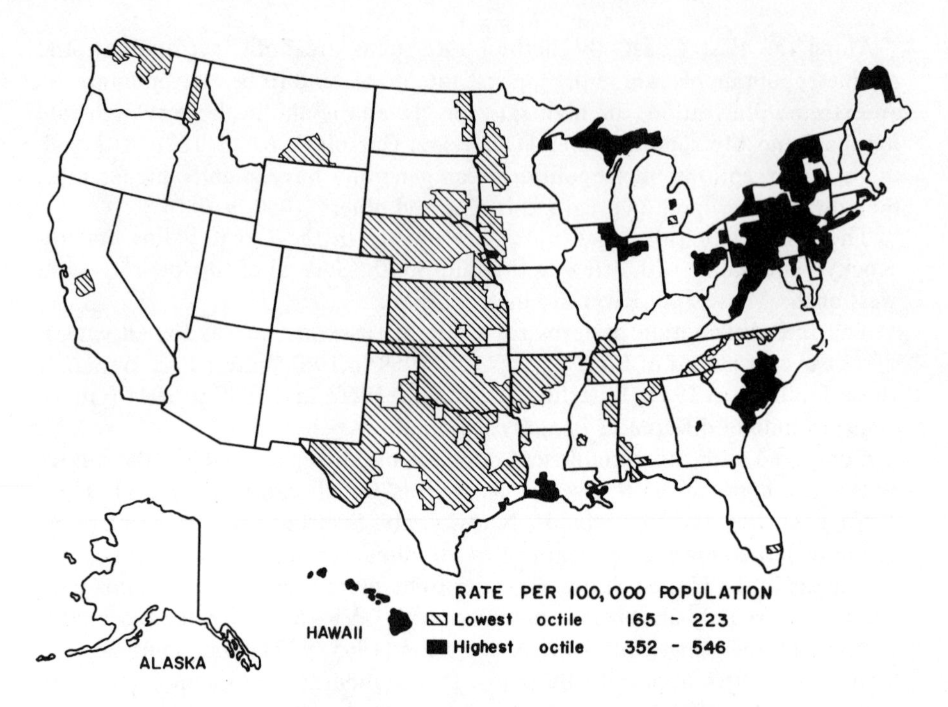

Figure 4. CVR death rates, white females, age 45 to 64, 1959–1961.

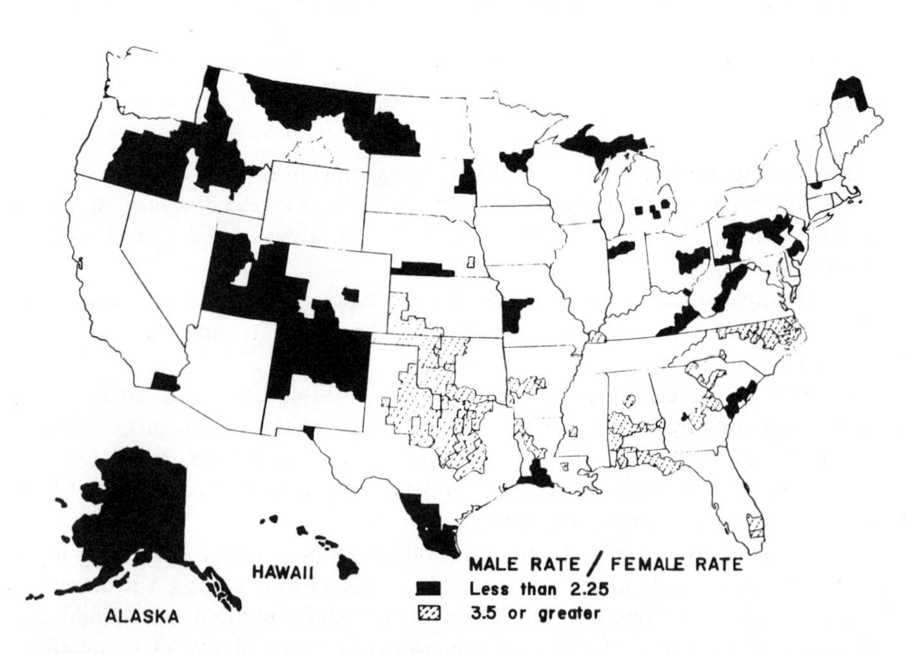

Figure 5. State economic areas with the highest and lowest ratios of death rates for the cardiovascular-renal diseases, whites, age 45 to 64, 1959–1961.

TABLE 1. DEATH RATES FOR ALL CAUSES, CVR DISEASES, AND MALIGNANT NEOPLASMS, BY SEX, AGE 45–64, SELECTED COUNTRIES, 1959–1961*

Country	Male			Female		
	All causes	CVR diseases†	Malignant neoplasms‡	All causes	CVR diseases†	Malignant neoplasms‡
United States (white population)	1473.7	813.8	286.7	719.6	308.5	241.8
Finland	1711.3	821.8	406.4	762.6	361.5	219.9
South Africa (white population)	1667.8	868.7	298.2	910.8	431.1	247.2
Austria	1431.5	518.4	375.3	747.7	257.1	281.2
France	1371.6	373.6	337.8	665.5	187.8	216.4
United Kingdom	1350.7	608.5	374.1	711.5	283.2	257.5
Australia	1343.9	754.6	245.3	721.8	352.5	202.1
Czechoslovakia	1322.8	490.6	385.0	725.4	262.0	257.9
German Federal Republic	1306.7	476.4	316.9	716.4	234.2	258.9
Canada	1263.0	696.6	256.7	697.5	298.4	246.0
Hungary	1261.3	467.3	309.5	788.4	326.5	251.9
New Zealand	1214.9	659.7	254.9	723.7	312.0	240.3
Italy	1198.6	434.8	306.6	660.1	267.1	214.6
Switzerland	1138.1	418.0	306.9	654.1	226.2	236.3
Denmark	1004.5	448.7	288.4	665.3	198.7	286.3
Israel (Jewish population)	973.9	537.7	209.9	730.1	306.7	250.6
Netherlands	971.5	365.3	313.2	567.6	173.2	244.0
Norway	935.8	430.7	225.2	537.5	194.2	219.3
Sweden	898.5	420.9	211.2	598.7	222.1	236.9

*Average annual death rates per 100,000 population, age-adjusted by direct method by 10-year age groups to total U.S. 1950 population in those age groups.
†Cardiovascular-renal diseases, International List nos. 330–334, 400–468, and 592–594.
‡Malignant neoplasms, International List nos. 140–205.
Source: World Health Organization (1962–1965).

Sex Ratios

For whites aged 45 to 64, CVR diseases rates for males are about three times those for females. Utah areas are among those with low ratios, with male rates only about twice the female rates (Fig. 5). This state has the lowest per capita sale of cigarettes (Friedman, 1967, p. 771); moreover, the Mormon Church forbids smoking. Males have generally smoked more than females (Haenszel and others, 1956); therefore, the hypothesis may be proposed that the low male rates and low sex ratios in Utah may be due to lower levels of cigarette consumption. Whether this hypothesis is correct or not, it illustrates a way in which a sex-related risk factor may be studied.

Some of the highest sex ratios are observed for portions of Oklahoma and central Texas with white male rates approximately four times the white female rates. Thus, as we test hypotheses regarding low rates in the Great

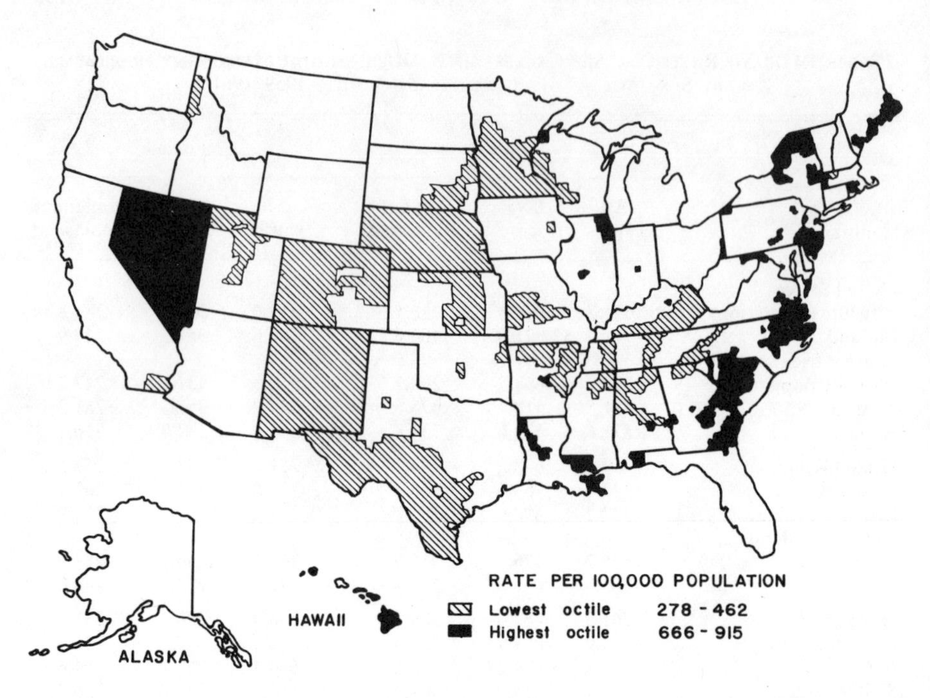

Figure 6. Coronary heart disease (cause 420) death rates, white males, age 45 to 64, 1959–1961.

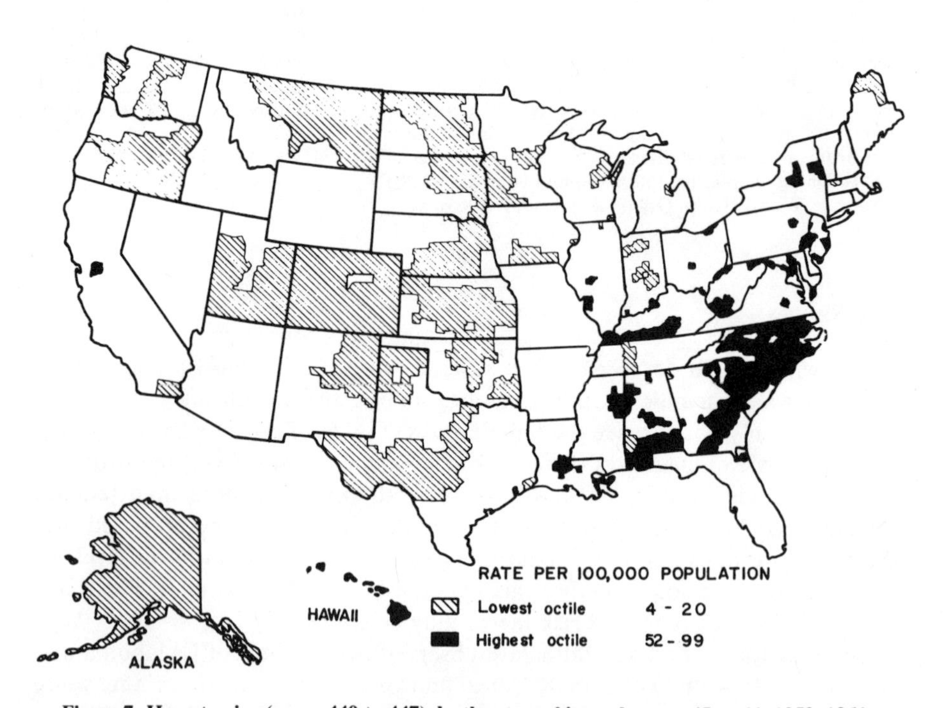

Figure 7. Hypertension (cause 440 to 447) death rates, white males, age 45 to 64, 1959–1961.

Plains, it is necessary to recognize that the Oklahoma-Texas areas have the lowest rates for females but only moderately low rates for males. This sex difference provides an additional resource for generating and testing hypotheses.

Low sex ratios are observed for New Mexico and some Texas areas, but an explanation is apparent: Mexican-born males have the lowest CVR-diseases death rates of white ethnic groups in the United States, while the Mexican-born females have one of the highest rates, resulting in a very low sex ratio of rates. We do not know whether this special pattern is due to genetic, cultural, or other factors. Further, we do not know the rates for whites other than those of Latin American ancestry. Thus, if the environment is contributing to the favorable white female experience of Oklahoma-Texas areas, it is possible that it is also present and contributing in New Mexico. In other words, we are pointing to the need, when testing any specific hypothesis (such as to the effect of geochemistry upon the risk of dying), to control not only for age, sex, and race, but also for other pertinent factors—in this instance, ethnic background.

SPECIFIC CVR CATEGORIES

Coronary Heart Disease

Coronary heart disease is the leading cause of death, and when comparable rates of death from this condition are determined by areas, they present a geographic pattern that is generally similar to that for all CVR diseases (Fig. 6). The Mexican-born males have low coronary-heart-disease death rates, as do many of the areas in which there are large numbers of men of Mexican ancestry, even though in a few instances the all-causes rates are not low (National Center for Health Statistics, unpublished data).

Metropolitan areas generally have higher rates for coronary heart disease than do nonmetropolitan areas, but the highest octile includes 21 nonmetropolitan areas, chiefly areas near the East Coast.

Hypertension and Stroke

High blood pressure or *hypertension,* which, as a cause of death, is usually accompanied by *hypertensive heart disease,* presents a geographic pattern of rates for white males (Fig. 7) to a considerable extent similar to the pattern for the CVR diseases but with high rates particularly in the Southeast and in metropolitan areas east of the Mississippi River.

Vascular lesions of the central nervous system, which include both *cerebral hemorrhage* and *central thrombosis* (often called "stroke"), is the third leading cause of death for all ages. It presents a geographic pattern somewhat like that for hypertension yet different in several respects; the highest rate areas are concentrated to an even greater extent in the Southeast and nearby; fewer metropolitan areas are included in the highest rate octile, while more, especially in the Northeast, are in the lowest octile; the lowest rate octile includes more areas for stroke than for hypertensive disease in the Rocky

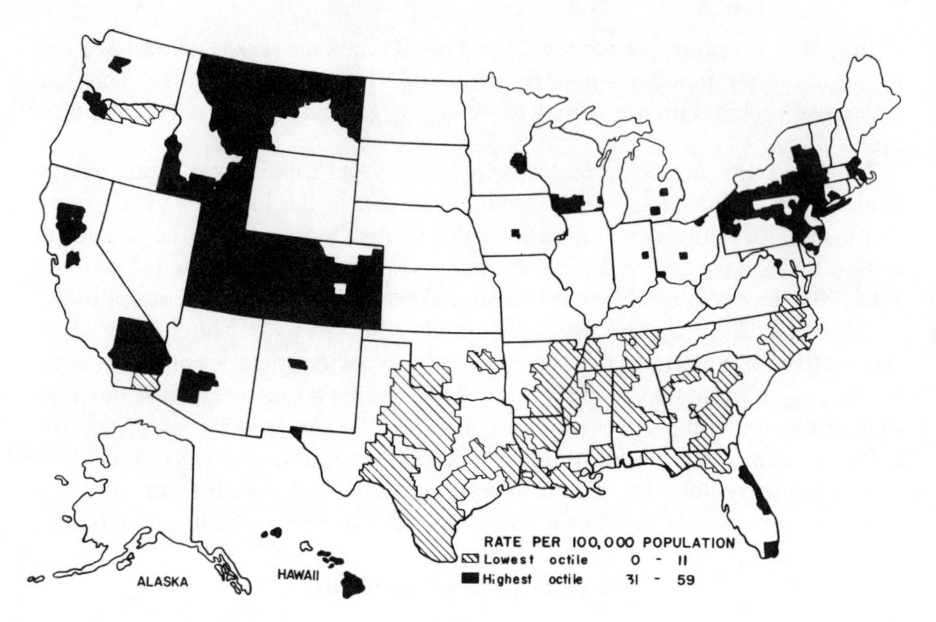

Figure 8. Rheumatic heart diseases (cause 400 to 416) death rates, whites, both sexes, age 45 to 74, 1959–1961.

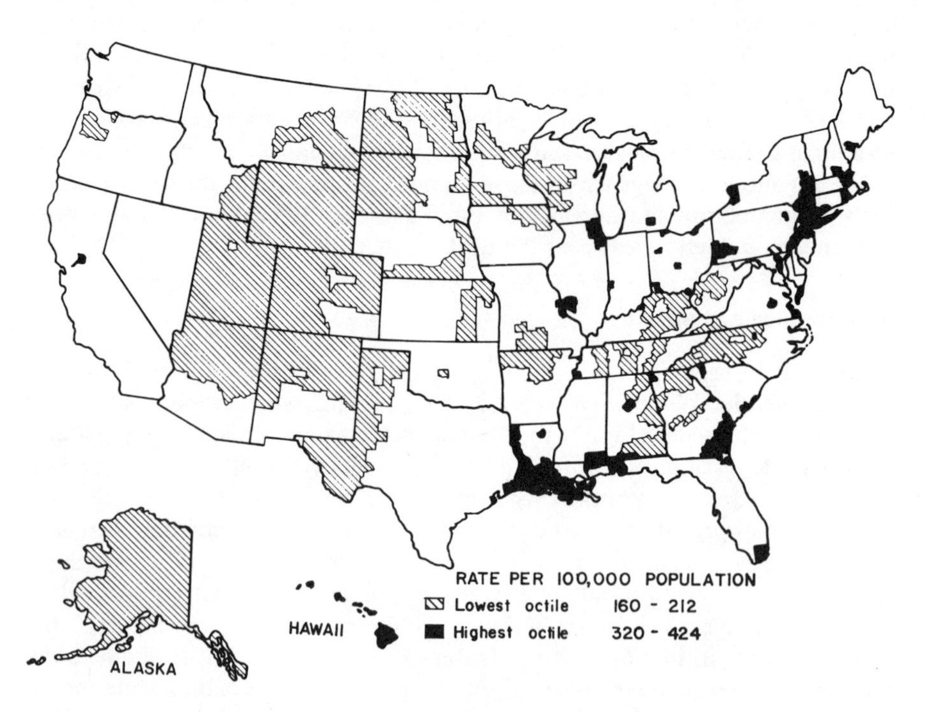

Figure 9. Malignant neoplasms (cause 140 to 205) death rates, white males, age 45 to 64, 1959–1961.

Mountains. A field study of six areas with marked contrasts in rates confirms the accuracy of the data as compared with other cause categories (Kuller and others, 1968).

Rheumatic Heart Disease

Rates for *rheumatic heart disease* and *rheumatic fever* are presented for whites of both sexes, age 45 to 74, age-sex-adjusted, in order to reduce the effect of chance fluctuation or random error (Fig. 8). Rates are generally lowest in the South, in conformity with clinical impressions and morbidity data (Marienfeld and others, 1964, p. 797; Perry and others, 1968, p. 923). Since a large proportion of the population of Florida moved to that state in adult life, and since the original attack often occurs in childhood, the high rates for two areas in that state are not necessarily related to the local environment. In general, the geographic pattern for this cause of death is not associated or correlated with that of any other cause category thus far studied.

MALIGNANT NEOPLASMS

Malignant neoplasms rank second as a cause of death. The highest death rate areas among middle-aged males are more consistently metropolitan, and the lowest rate areas are more consistently nonmetropolitan than are the rates for the CVR diseases (Fig. 9). There are also substantial similarities with the highest rate areas generally east of the Mississippi or along the Gulf Coast and the lowest rate areas in the Great Plains, the Rocky Mountains, and the mid-South. Thus, in spite of the greater concentration of high cancer rates in metropolitan areas, the correlation of cancer and CVR death rates for the 509 state economic areas is definitely positive ($r = +0.53$, $p<0.001$) and is consistently one of the highest observed in cause-correlation matrices.

United States male rates for all forms of cancer combined compare rather favorably with those for other countries (Table 1). Only one U.S. area has a higher rate than the Finnish rates while 54 areas have lower rates than the lowest rates for any entire country.

Lung-cancer rates have risen rapidly in recent years; lung cancer causes more deaths among middle-aged males than any other form of neoplasm, thus providing strong stimulus for studying this disease. The areas with the lowest rates are generally nonmetropolitan (Fig. 10), located in the Rocky Mountains and Great Plains, with some in the mid-South. The highest rate areas are nearly all metropolitan and in the eastern half of the country. The nonmetropolitan areas with high rates are mostly along or near the Gulf Coast.

Community air pollution presents an obvious hypothesis for the high urban rates for lung cancer. Individual air pollution by cigarette smoking presents a hypothesis for geographic variations in rates for lung cancer, for coronary heart disease, and for the association between these two categories. The category "all malignant neoplasms except lung cancer" also consistently shows

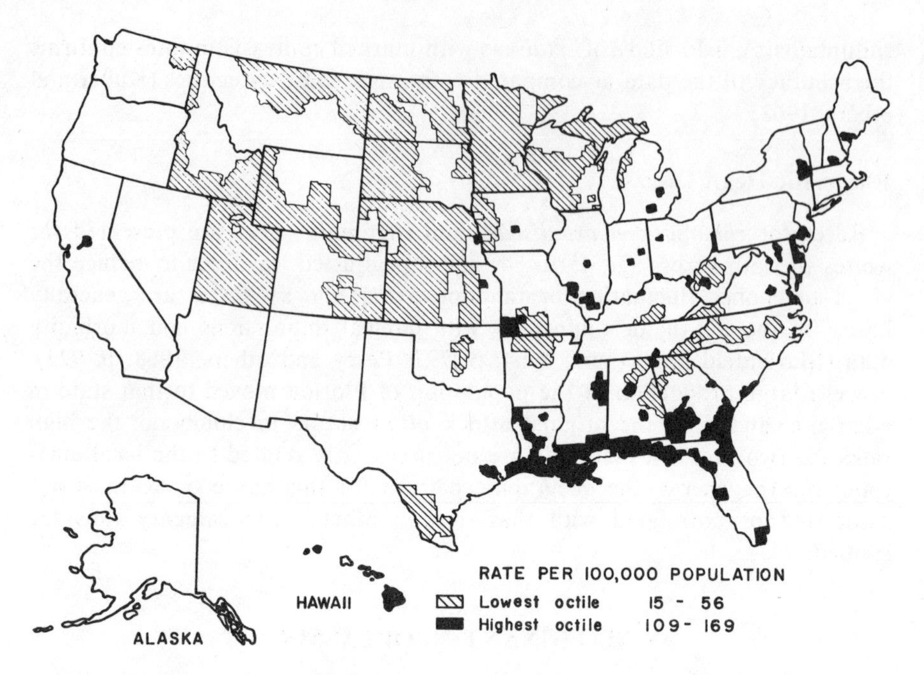

Figure 10. Lung cancer (cause 162, 163) death rates, white males, age 45 to 64, 1959 to 1961.

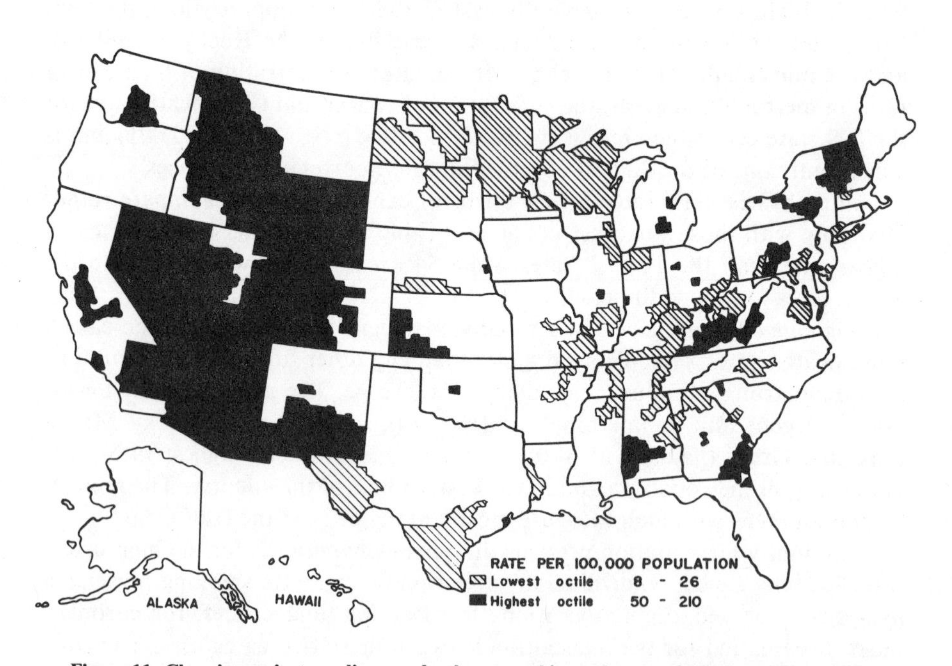

Figure 11. Chronic respiratory diseases death rates, white males, age 35 to 74, 1959 to 1961.

equally substantial "statistically significant" correlations with both coronary heart disease and the CVR diseases as a group.

OTHER CAUSES

Death rates for *chronic respiratory diseases*—principally *chronic bronchitis, emphysema,* and *asthma,* have also increased rapidly in recent years. The geographic patterns are independent of other cause categories, however (Fig. 11). The extremely high rates for Arizona are due primarily to the high rates for those moving to Arizona, not to those born there (Sauer, 1967, p. 405).

For the category chronic respiratory diseases, the number of deaths is small enough, for age 45 to 64 and also for 35 to 74 (age-adjusted), to allow chance fluctuation to play an important role in identifying some of the areas with outstanding rates. Rates for the broader age group for the larger states, however, also show a pattern essentially independent of the patterns of mortality for either coronary heart disease or lung cancer—or other disease categories for this study. In view of the very substantial amount of evidence regarding the role of cigarette smoking in all three of these diseases, further work is needed to understand the geographic patterns of chronic respiratory disease.

Accidental and violent causes, as a group, also fluctuate independently of other causes, as might be expected if we assume that their etiology tends to be independent of the etiology of the various "natural" causes of death.

All other causes, which include *pneumonia, diabetes, cirrhosis of the liver,* and *all remaining causes* of death, as a group, show a moderate degree of association with the CVR diseases regarding geographic patterns.

"LATE MIDDLE AGE"

Other approaches in epidemiological and public-health statistics may be studied, in some instances for adding specificity to mortality patterns and in other instances for identifying limitations in the usefulness of the data in the search for environmental factors causing differences in mortality patterns. As an example, the death rates for age 65 to 74 or "late middle age" are generally similar to those for age 45 to 64, but with a major exception—the low rates for Florida areas (Fig. 12). Death rates by state of birth show clearly that, for this age group, Florida residents who were born there have roughly average rates, although those moving to Florida have low rates (Sauer, 1967, p. 404). This may be due to the self-selection of migrants to Florida who have a higher potential for longevity.

POTENTIAL BENEFITS

We may raise the questions: What is the potential importance of these geographic differences in reducing the number of deaths in the United States under age 65? If the death rates of the lowest rate areas applied for the United States as a whole, how many fewer deaths per year would there be?

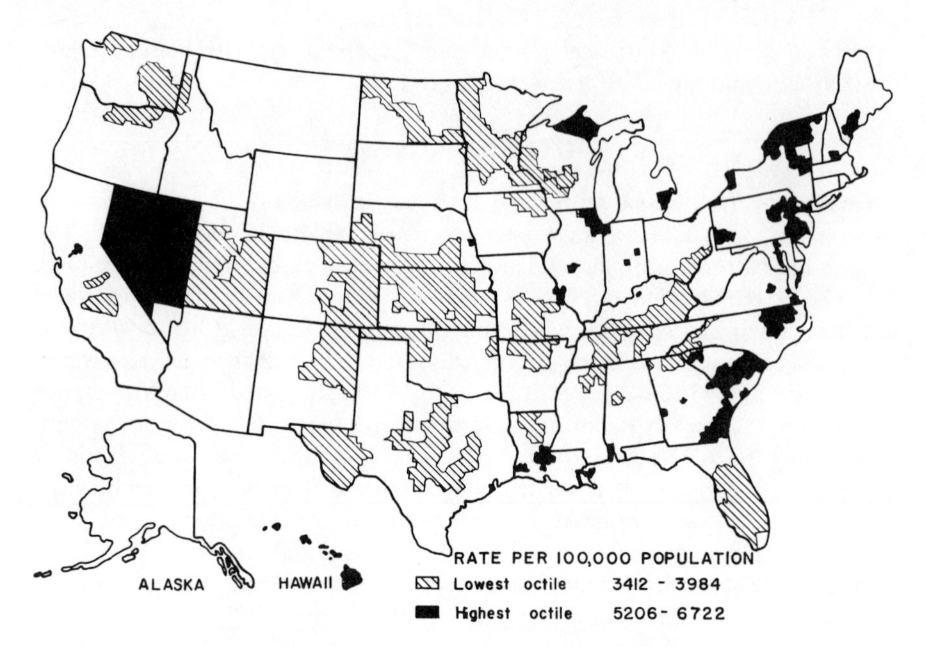

Figure 12. All-causes death rates, white males, age 65 to 74, 1959 to 1961.

Areas were determined on the basis of the 1949 to 1951 rates for all causes of death, white males aged 45 to 64. Four economic subregions were selected on the basis of their 1949 to 1951 death rates, but their 1959 to 1961 rates were used in the calculation, in order to hold random error to a minimum. These economic subregions are located in central Minnesota, west-central Minnesota, and northeastern South Dakota; west-central Tennessee and northeastern Mississippi; northern Kansas and south-central Nebraska. They consist of 9 SEA's, or 82 counties, and have a population of more than 1 million whites.

If the rates of the four lowest subregions replaced the actual U.S. rates, the reduction in deaths per year would be:

under 35 years	2,051
35 to 44	7,716
45 to 54	29,146
55 to 64	57,982
Total	96,985

These figures would be even larger if nonwhites were included in the analysis. Each age-sex group in the low-rate areas had rates lower than the U.S. rate and thus contributed to a potential reduction of almost 100,000 deaths per year under age 65. If the rates for Nebraska area 5, the lowest death-rate area, applied to the entire United States, the total annual reduction would be 131,634 under age 65 alone.

This potential saving in lives in middle age and earlier years cannot be predicted with precision. If portions of these differences are due to factors that cannot in practice be changed, then the potential saving would be less. If, on the other hand, the pertinent etiological factors in the low-rate areas are not present in optimum amounts, the potential saving may be greater. In view of the number of entire European countries with lower rates than any of our lowest rate areas, this latter alternative should not be lightly dismissed.

DISCUSSION

To achieve this potential saving in lives that we have discussed, it is necessary to recognize the nature of the geographic patterns. The similarities of various cause categories have been great enough to raise the question as to whether or not it is feasible to study the geographic patterns of any one cause separately (Syme, 1968; Sauer, 1962, p. 94). In addition to a study of conventional grouping of causes, such as CVR diseases, other possible combinations of categories may be considered. An alternate approach, which may be used simultaneously, is to endeavor to define lethal entities with more precision or in a different manner than is now done. Present attempts in this direction merely suggest the need for further study.

Many hypotheses have been proposed as to factors responsible for geographic differences in these various cause-specific death rates (Schroeder, 1960, 1966; Sauer, 1962, p. 101–103; Buechley and others, unpub. data; Dudley and others, 1969). Chemical characteristics of the drinking water, other geochemical factors, and many other factors in the physical and cultural environment have been demonstrated to have moderate and "statistically significant" degrees of association with geographic patterns of death rates, but some seem to be most unlikely candidates as etiological factors. Further study is in order to determine which inferences may be properly drawn from such correlations (Duncan and others, 1961, p. 10, 106*ff*). Some correlations or associations are coincidental, a result of special distributions or factors inadequately perceived by the investigator, technically classified as "spurious"; others may result from both variables having a common cause; still others may be causal (Buechley and others, 1966, unpublished).

Simultaneous recognition of both similarities (or associations) and differences in geographic patterns according to cause, sex, and other factors provides a framework for more specific testing of the many hypotheses in the environment that in some degree may be related to the risk of dying. Massive quantities of data already collected provide a means of testing some hypotheses, as a complement to intensive studies in a few areas. At present, little is available as definitive answers, but progress is being made in increasing the precision as well as the number of questions being raised, and in forms that facillitate further study.

Death constitutes the most objective, most easily measured, and usually the most serious effect of the environment upon a population. Furthermore, pertinent data are available from well-established vital-record systems. For some diseases such as rheumatic fever and rheumatic heart disease, which may have their initial onset decades prior to death, caution may appropriately be exercised in the interpretation of mortality data; under such circumstances, disease incidence, or the number of new cases developing per year, is likely to be the more appropriate measure, in spite of the substantial difficulties in obtaining such data. Disease prevalence is less likely to be useful, as a measure of factors causing a disease, but may in some instances be useful in measuring the effect of the environment upon survival or recovery, even though adequate prevalence data are likely to be even more difficult to obtain than incidence data. Other levels of evaluation suggested by White (1967, p. 849), disability, discomfort, and dissatisfaction, also may be considered, along with the problems of arriving at feasible operational definitions of these terms, in the further study of areas with unusual patterns of mortality.

In summary, marked geographic differences in death rates have been presented for middle-aged whites in the United States. Factors responsible for these differences have not yet been clearly defined. Low death rates prevailed in the Great Plains in both 1959 to 1961 and 1949 to 1951. Contrastingly, high rates (sometimes double the low rate) are dominant along the East Coast. These patterns hold true basically for both males and females, but with the lowest female rates concentrated in Oklahoma and central Texas.

Geographic variations in death rates also exist among different countries. Several foreign countries have rates for middle-aged males definitely lower than the rates for any of the 509 state economic areas in the United States.

Many factors are currently being studied for their potential contribution to the geographic differences in the risk of dying in middle age. It seems reasonable to predict that a number of factors will be found to contribute to these differences, some of which may currently not even be suspected.

REFERENCES CITED

American Heart Association, National Heart Institute, and Heart Disease Control Program, 1965, Cardiovascular Diseases in the U.S., facts and figures: New York, American Heart Association.

Buechley, R. W., Oechsli, F. W., and Stallones, R. A., unpublished, 1966, What Meanings Can Be Read into Geographic Variations in Cardiovascular Mortality?

Dudley, E. F., Beldin, R. A., and Johnson, B. C., 1969, Climate, water hardness and coronary heart disease: Jour. Chron. Diseases, v. 22, p. 25–48.

Duffy, E. A., and Carroll, R. E., 1967, United States metropolitan mortality, 1959–1961: PHS Pub. no. 999-AP-39, U.S. Public Health Service, National Center for Air Pollution Control.

Duncan, O. T., Cuzzort, R. P., and Duncan, B., 1961, Statistical geography: The Free Press of Glencoe, Illinois, 191 p.

Enterline, P. E., and Stewart, W. H., 1956, Geographic patterns in death from coronary heart disease: Public Health Rept., v. 71, p. 849–855.

Enterline, P. E., Rikli, A. E., Sauer, H. I., and Hyman, M., 1960, Death rates for coronary heart disease in metropolitan and other areas: Public Health Rept., v. 75, no. 8, p. 759–766.

Friedman, G. D., 1967, Cigarette smoking and geographic variations in coronary heart disease mortality in the United States: Jour. Chron. Diseases, v. 20, p. 769–779.

Gover, M., 1949, Statistical studies of heart disease, IV. Mortality from heart disease (all forms) related to geographic section and size of city: Public Health Rept., v. 64, no. 14, p. 439–456.

Haenszel, W., Shimkin, M. B., and Miller, H. P., 1956, Tobacco smoking patterns in the United States: Public Health Service Pub. 463, U.S. Dept. Health, Education, and Welfare.

Kuller, L. H., Bolker, A., Saslaw, M. S., Paegel, B. L., Sisk, C., Borhani, N., Wray, J., Anderson, H., Peterson, D., Winkelstein, W., Jr., Cassel, J., Spiers, P., Robinson, A. G., Curry, H., Lilienfeld, A. M., and Seltser, R., 1968, Nationwide cerebrovascular disease mortality study: Presented at the American Heart Association meeting, Miami, Florida, November 24, 6. p.

Linder, F. E., and Grove, R. D., 1943, Vital statistics rates in the United States, 1900–1940: Washington, U.S. Bureau of the Census, Vital Statistics, 1051 p.

Marienfeld, C. J., Robins, M., Sandidge, R. P., and Findlan, C., 1964, Rheumatic fever and rheumatic heart disease among U.S. college freshman, 1956–60: Public Health Rept., v. 79, no. 9, p. 789–811.

National Center for Health Statistics, 1949–1966, Vital statistics of the United States: Annual volumes.

——, 1968, Comparability of age on the death certificate and matching census record, United States, May–August, 1960: Vital and Health Statistics, PHS Pub. no. 1000, ser. 2, no. 29, Washington.

Perry, L. W., Poitras, J., and Findlan, C., 1968, Rheumatic fever and rheumatic heart disease among U.S. college freshman, 1956–65: Public Health Rept., v. 83, no. 11, p. 919–938.

Sauer, H. I., 1962, Epidemiology of cardiovascular mortality—geographic and ethnic: Am. Jour. Public Health, v 52, p. 94–105.

——, 1966, Adequacy of age data for age-specific death rates: reliability and validity of age as entered on death certificates, Charleston County, South Carolina, 1961–1963: Presented at Population Association of America, April 30.

——, 1967, Migration and the risk of dying: American Statistical Association Proceedings of the Social Statistics Section, p. 399–407.

Sauer, H. I., and Enterline, P. E., 1959, Are geographic variations in death rates for the cardiovascular diseases real?: Jour. Chron. Diseases, v. 10, p. 513–524.

Sauer, H. I., Payne, G. H., Council, C. R., and Terrell, J. C., 1966, Cardiovascular disease mortality patterns in Georgia and North Carolina: Public Health Rept., v. 81, no. 5, p. 455–465.

Schroeder, H. A., 1960, Relation between mortality from cardiovascular disease and treated water supplies: Am. Med. Assoc. Jour., v. 172, p. 1902–1908.

——, 1966, Municipal drinking water and cardiovascular death rates: Am. Med. Assoc. Jour., v. 195, no. 2, p. 81–85.

Stewart, W. H., and Enterline, P. E., 1957, Ecology and coronary heart disease: Jour. Chron. Diseases, v. 6, p. 86–89.

Syme, S. L., 1968, Is there a future in the epidemiological study of coronary heart disease?: Presented at Am. Public Health Association, Detroit, November 11.

U.S. Bureau of the Census, 1963, U.S. census of population: 1960 subject reports, State Economic Areas: PC (3)—1A, Washington, U.S. Government Printing Office.

——, 1964a, Evaluation and research program of the U.S. censuses of population and housing, 1960: Accuracy of data on population characteristics as measured by reinterviews: series ER60, no. 4: Washington, U.S. Government Printing Office.

U.S. Bureau of the Census, 1964b, Accuracy of data on population characteristics as measured by CPS—Census match: series ER60, no. 5: Washington, U.S. Government Printing Office.

White, K. L., 1967, Improved medical care statistics and the health services systems: Public Health Rept., v. 82, p. 847–854.

World Health Organization, 1957, International classification of diseases, seventh revision conference, 1955: Geneva, World Health Organization.

——, 1962, 1963, 1964, Annual epidemiological and vital statistics, 1959, 1960, and 1961: Geneva, World Health Organization.

——, 1965, World health statistics annual, 1962, v. 1: Geneva, World Health Organization.

MANUSCRIPT RECEIVED BY THE SOCIETY JULY 18, 1969

Printed in U.S.A.

THE GEOLOGICAL SOCIETY OF AMERICA, INC.
MEMOIR 123, 1971

Calcium-Carbonate Hardness of Public Water Supplies in the Conterminous United States

BARBARA M. ANDERSON
U.S. Geological Survey, Denver, Colorado

ABSTRACT

Geographic variations in the hardness of water seem to parallel the geographic variations in certain cardiovascular diseases (Anderson and others, 1969; Biorck and others, 1965; Schroeder, 1966). Winton and McCabe (1970) have provided a recent review of the problem. The map, presented here, showing calcium-carbonate hardness of municipal raw-water sources in the conterminous United States has been compiled to provide data for further examination of correlations between water hardness and cardiovascular diseases.

CONTENTS

DISTRIBUTION OF WATER HARDNESS

Water hardness is popularly recognized as the property of water that prevents or minimizes lathering with soap. In the past, hardness was measured by the actual amount of soap needed to make a lather (Durfor and Becker, 1964, p. 23). Hardness of water now, by convention, is expressed in terms of an equivalent quantity of calcium carbonate based on chemical analysis for calcium, magnesium, carbonate, and bicarbonate ions (Hem, 1959, p. 34). Other elements that contribute to hardness of water are aluminum, iron, manganese, strontium, and zinc, as well as compounds such as free acid (Lohr and Love, 1954a, and 1954b, p. 12).

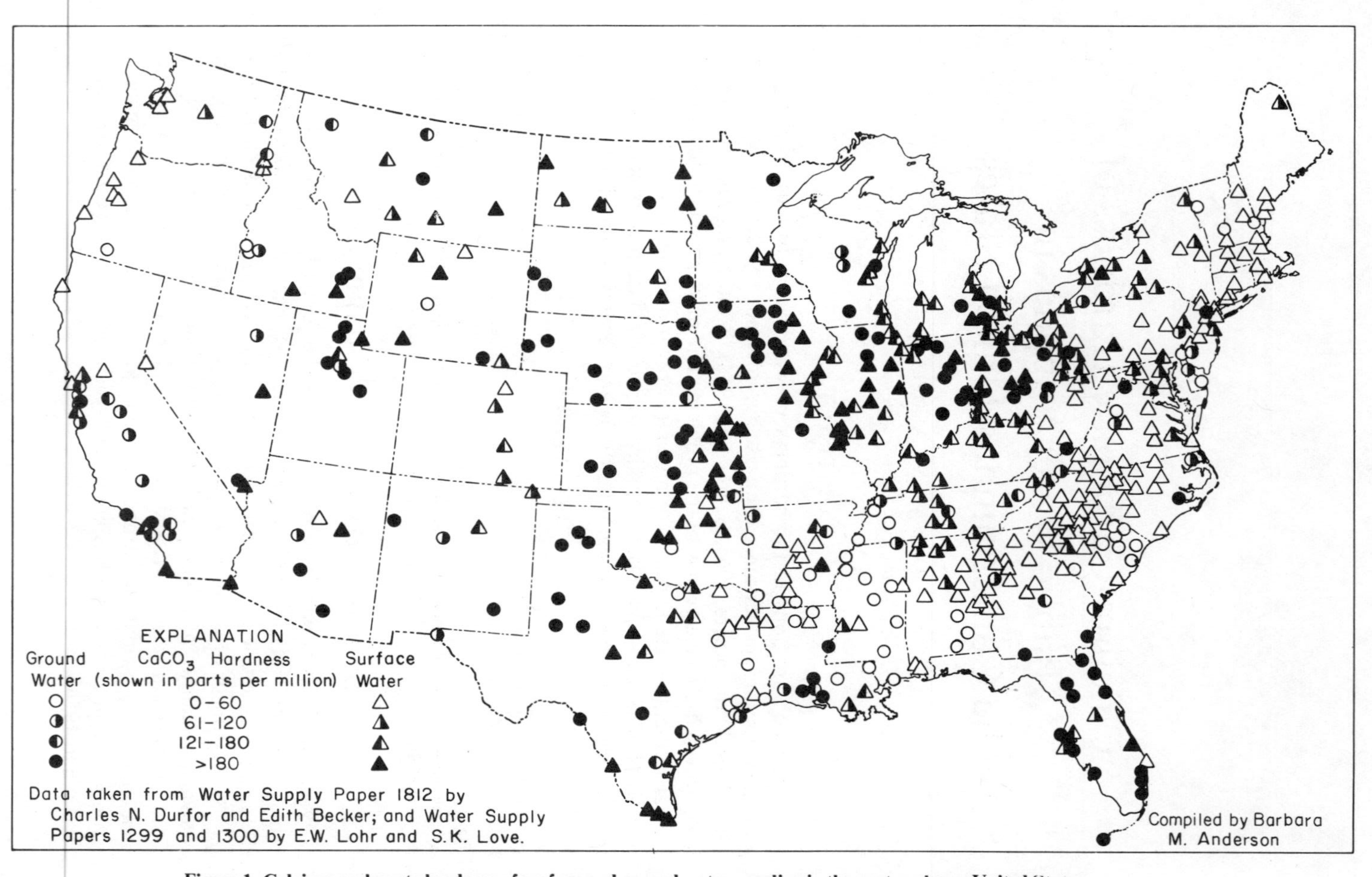

Figure 1. Calcium-carbonate hardness of surface and ground water supplies in the conterminous United States.

There is no clear line of demarcation between soft and hard water; consequently, hardness of water is shown on the map (Fig. 1) by symbols that represent ranges of amounts of equivalent calcium carbonate in parts per million. The softest waters have hardnesses of 0–60 ppm equivalent calcium carbonate, and the hardest waters have at least 180 ppm equivalent calcium carbonate (Durfor and Becker, 1964, p. 27).

Data on calcium-carbonate hardness of raw, untreated water supplies were compiled from Lohr and Love (1954a, 1954b) and Durfor and Becker (1964), and consist of yearly averages as given in those reports. The analyses were separated into two groups depending on the source of the water: (1) ground waters—springs, wells, and infiltration galleries; and (2) surface waters—rivers, lakes, and reservoirs. Some localities utilize both ground- and surface-water sources; and those localities have been classified according to the predominant source. Hardness of the water reflects the complexity of geologic factors that affect water composition (Hem, 1959, p. 201–216). The data do not represent the hardness of the treated water actually utilized by the population at the localities shown. Treatment, where practiced, may result in water either softer or harder than the raw supply.

REFERENCES CITED

Anderson, T. W., le Riche, W. H., and Mackay, J. S., 1969, Sudden death and ischemic heart disease—Correlation with hardness of local water supply: New England Jour. Medicine, v. 280, no. 15, p. 805–807.

Biorck, Gunnar, Bostrom, Harry and Widstrom, Anders, 1965, On the relationship between water hardness and death rate in cardiovascular diseases: Acta Medica Scandinavica, v. 178, no. 2, p. 239–252.

Durfor, Charles N., and Becker, Edith, 1964, Public water supplies of the 100 largest cities in the United States, 1962: U.S. Geol. Survey Water-Supply Paper 1812, 364 p.

Hem, John D., 1959, Study and interpretation of the chemical characteristics of natural water: U.S. Geol. Survey Water-Supply Paper 1473, 269 p.

Lohr, E. W., and Love, S. K., 1954a, States east of the Mississippi River, Pt. 1 *of* The industrial utility of public water supplies in the United States, 1952: U.S. Geol. Survey Water-Supply Paper 1299, 639 p.

——, 1954b, States west of the Mississippi River, Pt. 2 *of* The industrial utility of public water supplies in the United States, 1952: U.S. Geol. Survey Water-Supply Paper 1300, 462 p.

Schroeder, H. A., 1966, Municipal drinking water and cardiovascular death rates: Am. Medical Assoc. Jour., v. 195, no. 2, p. 125–129.

Winton, Elliott F., and McCabe, Leland J., 1970, Studies relating to water mineralization and health: Am. Water Works Assoc Jour., v. 62, no. 96, p. 26–30.

PUBLICATION AUTHORIZED BY THE DIRECTOR, U.S. GEOLOGICAL SURVEY, WASHINGTON, D.C.

MANUSCRIPT RECEIVED BY THE SOCIETY APRIL 13, 1970

Printed in U.S.A.

THE GEOLOGICAL SOCIETY OF AMERICA, INC.
MEMOIR 123, 1971

The Geochemist's Involvement with the Pollution Problem

HELEN L. CANNON
AND
BARBARA M. ANDERSON
U.S. Geological Survey, Denver, Colorado

ABSTRACT

The geochemist can contribute much information of value toward assessing the effect of environment, including inorganic pollution, on health. The average composition of rocks, soils, plants, and water and also the increments of inorganic substances that can be expected in geologic environments of high-metal content are essential for comparison with metal contents of these components of the environment in areas contaminated by various types of inorganic air and water pollution.

Background levels of lead, zinc, nickel, chromium, copper, and manganese in soils and in four classes of vegetation have been estimated from collections that were made in remote areas presumed to be free from inorganic contamination. The trace-metal content of soils and plants varies widely in different geologic provinces of the United States; in those areas of high natural mineralization, additions of metals from man-made pollution may compound a possible hazard. Results of sampling in urban areas show that contamination of vegetation by gasoline lead can be expected for at least 1000 ft back from transportation lanes, and that the lead burden is increasing greatly with time. Ore-treatment plants can also be a source of air contamination for several miles downwind and several thousand feet in other directions. Soils of naturally high metal content in a mining district may be further contaminated with both major and minor elements from smelting operations. Recent studies show that volatile elements are released directly to the air from ore deposits in place. Concentrations of mercury, for instance, may be as much as 20 times background for several hundred feet in altitude over ore deposits in which mercury occurs as a relatively minor constituent.

The source of inorganic pollution in surface drainage and also in ground water is commonly geologic, as rivers may be contaminated from coal and metal deposits in place and also from mining and smelting operations.

More information of the type illustrated should be accumulated and made available to scientists who are working in environmental health, and, in particular, to those involved in the pollution problem. Only by these means can we provide a scientific basis for the enactment of realistic and effective legislation for pollution control.

CONTENTS

INTRODUCTION

My privy and well drain into each other
After the custom of Christendie
Fever and fluxes are wasting my mother
Why has the Lord afflicted me?

As was the sowing so the reaping
Is now and evermore shall be.
Thou art delivered to thine own keeping.
Only thyself can deliver thee!

RUDYARD KIPLING
1914

The dangers from epidemics of transmissible bacterial diseases have been decreased markedly through improved sanitation practices since Rudyard Kipling wrote these words, but the problem of waste disposal keeps apace with population and industry—ever on the increase. Those who are studying the effects of pollution in the world today have been concerned largely with man's contribution to air and water quality and have overlooked the broader meaning of the term, which includes impurity that is a consequence of natural phenomena; for example, anomalous distribution of elements in the substrate. The increments of metal contamination that have been and are being added to these naturally anomalous areas by mining and ore treatment practices are also large.

Many of the inorganic constituents of our physical environment are of geologic origin, or at least occur in the weathering cycle with which the geochemist is familiar. Kinds of information that the geochemist can contribute and that should be valuable in studies of environmental health, and particularly in the pollution problem are: (1) the average background amounts of each element to be expected in a natural uncontaminated geologic setting; (2) the effects of air pollution on the background levels of elements in soils and plants; (3) the geochemistry of soils and the solution and migration of elemental components through surface and ground waters; (4) the chemical composition of mining and milling wastes and the solubility of the elemental components, which result in either their retention at the site or their dispersion through local drainage systems; (5) the accumulation by vegetation of salts and toxic elements in surface soils; (6) the formation and cyclical deposition of natural evaporative salts and volatile elements of geologic origin.

Before we set standards for optimum chemical quality that can serve as a valid basis for legislation to control the pollution of our physical environment, many geochemical data must be accumulated and interpreted. Some examples of the range in metal contents in plants, soils, and waters of natural and of polluted environments are presented.

The samples, unless otherwise specified, were collected by the authors and analyzed in the U.S. Geological Survey laboratories. Contents of metals were measured in ash and later calculated in the dry weight, except where otherwise noted.

BACKGROUND LEVELS OF METALS IN UNCONTAMINATED SOILS AND VEGETATION

The detection of trace metals in plants and soils and the establishment of background values that may be expected in an unpolluted natural geologic setting are important to the geochemist in his study of the dispersion of metals in the weathering cycle.

Because of the daily increase of pollution in our atmosphere, it is important to establish background levels in the vegetation and soils from the remote areas of our country that are relatively free from contamination. From remote areas in several geologic settings, 67 vegetation and 39 soil samples were collected to obtain a general average. The mean and median contents of *lead, chromium, nickel, copper, manganese,* and *zinc*[1] obtained in ash and calculated in dry weight of the samples are given in Table 1. As can be seen, the uptake of metals by the various major classes of plants differs considerably. In general, grasses absorb less metal than other plants, and trees absorb much zinc and manganese. Cadmium was not included in Table 1, because a good analytical method for determining its content was not at that time available. However, recent analyses indicate that the average cadmium content in the ash of plants is probably about 1 ppm (part per million) and that of soils a little less than 1 ppm. The average lead content (48 ppm) of uncontaminated plants collected in remote areas more than 1000 ft from roads was considerably less than that of plant samples that were collected within 300 ft from paved roads at approximately 750 sites in the United States (H. T. Shacklette, 1970, oral commun.). The lead contents (20 ppm) of B-horizon soils (soil from depth of 2–8 in.) collected from remote areas, on the other hand, were equal to those of 863 soils collected at the same depth within a few hundred feet from a paved road (Shacklette and others, 1970).

ANOMALOUS METAL CONTENTS IN NATURAL SOILS AND VEGETATION

There are many geologic areas of high-metal content, where crops, if grown, might contain amounts of metal above those limits set by the Food and Drug Administration as hazardous to health. The toxic effects of naturally occurring selenium in plants that grow on Cretaceous shales of the west are well documented (Trelease and Beath, 1949). Native vegetation in the closed basins of the Basin and Range province commonly contains large quantities of strontium, fluorine, boron, or lithium. Several hundred parts per million of lead, zinc, or copper, are commonly found in the ash of wild blueberries, raspberries, and other edible fruits that grow near deposits of these metals. The possibility of anomalous cadmium contents in plants that grow in lead-zinc districts is currently being investigated. The metal contents of plants that grow near deposits of mercury, chromium, cobalt, nickel, or other metals may also be high.

Anomalous amounts of metal occur in vegetables that are raised commercially on some drained peat areas. For example, metals like *zinc* that are removed from the ground water and adsorbed on the peat may become readily available to plants when the peat is drained and oxidized; other metals like copper may become unavailable and may actually be deficient in the crops. In one area of western New York, where drained peat bogs overlie a mineralized bed of Lockport Dolomite, plants growing on them were collected

[1]Because the spectrographic method for zinc analysis has a detection limit of 500 ppm and thus is unsuited for most plant samples, only those samples that had been analyzed chemically were used in the calculation.

in 1946, shortly after the bogs were drained, and were found to contain as much as 1 percent zinc in the dry weight (Cannon, 1955). In the last 20 years, the content of zinc in the vegetation has lessened considerably, as the oxidized zinc gradually has been removed from the peat by circulating ground water; no more than 490 ppm zinc is now found in samples of vegetables grown on the peat. Because a good analytical method for determination of *cadmium* content is currently available, peat samples from the area in New York recently have been analyzed for cadmium content and have been found to contain more than 100 ppm in the dry weight; crops contained as much as 8 ppm cadmium in the ash or 1 ppm in the dry weight. Whether the original cadmium contents were greater is not known, because the samples taken in 1946 were not analyzed for cadmium.

If plants absorb anomalous amounts of metal from geologic sources in some areas and, in addition, receive increments of metal from various sources of air and water pollution, as discussed in the next section, the possible health hazard for those areas would be considerably increased.

AIR POLLUTION

Urban Contamination

The burning of coal for heat has long been a major source of air pollution, particularly in cities near soft-coal-producing areas. Although the trend for cleaner heat by gas and electricity and the ordinances by cities against soft coal have lessened this type of air pollution considerably, electrical power plants are still emitting large amounts of coal products to the atmosphere of our cities. The *sulfur* content of the coal is important for its effect on smog quality. Dunn and Bloxam (1932, 1933) have reported the death of livestock pastured near coke ovens in England and were able to show, by analyses of soil, grass, and stack soot, that copper and manganese were being emitted from the ovens in significant quantities and that the amounts of *lead* ingested were sufficient to kill the stock.

Evidence of contamination of soils and vegetation from air pollution by vehicular traffic is slowly accumulating from various sources. Warren and Delavault (1960) have found as much as 3100 ppm *lead* in the ash of arborvitae twigs within 100 yds of a highway in Vancouver, Canada; the normal background is less than 50 ppm. Analyses of soils near highways in the Twin Cities Metropolitan area, Minneapolis and St. Paul, by Singer and Hanson (1969) showed abnormally high lead concentrations, with a maximum of 700 ppm on road embankments. The lead concentrations were found to be related to traffic volume and distance from highways, the levels being significantly higher within 15 m ($\approx$50 ft) of the highway. Lead concentrations of 50+ ppm at greater than 15 m were considered to be near normal.

Three transects were recently made across highways in Sweden by Rühling and Tyler (1968). They found that lead is concentrated in plants and soils near highways and that the values rapidly decrease to a normal content of less than

Table 1. Content, in Parts per Million, of Six Metals in Vegetation and Soils from Uncontaminated Areas of the United States

Type of vegetation or soil collected more than 1000 ft from nearest road	Number of samples*	Lead				Nickel				Chromium			
		Mean		Median		Mean		Median		Mean		Median	
		Ash	Dry wt.	Ash	Dry wt.	Ash	Dry wt.	Ash	Dry wt.	Ash	Dry wt.	Ash	Dry wt.
Vegetation													
Deciduous trees (leaves)	10	28	(1.8)	20	(1.3)	55	(3.5)	70	(4.5)	16	(1.0)	13	(0.8)
Grass (part above ground)	15	24	(1.9)	20	(1.6)	15	(1.0)	15	(1.2)	22	(1.8)	20	(1.6)
Shrubs (leaves and twigs)	22	45	(3.6)	30	(2.4)	35	(2.7)	30	(2.4)	46	(3.7)	30	(2.4)
Conifers (branch tips)	20	80	(2.8)	50	(1.8)	25	(.9)	15	(.5)	16	(.6)	15	(.5)
Average, all plants	67	48	(2.7)	30	(1.7)	30	(1.7)	20	(1.1)	27	(1.5)	20	(1.1)
Soils													
		Mean		Median		Mean		Median		Mean		Median	
"B" horizon	39	20		20		20		20		80		70	

Analyses are by semiquantitative spectrographic method, except where otherwise shown.

Table 1. (Continued)

Type of vegetation or soil collected more than 1000 ft from nearest road	Number of samples*	Zinc*				Copper				Manganese			
		Mean		Median		Mean		Median		Mean		Median	
		Ash	Dry wt.	Ash	Dry wt.	Ash	Dry wt.	Ash	Dry wt.	Ash	Dry wt.	Ash	Dry wt.
Vegetation													
Deciduous trees (leaves)	10	950	(60)	500	(32)	175	(11)	150	(10)	1600	(101)	850	(55)
Grass (part above ground)	15	140	(11)	150	(12)	100	(8)	100	(8)	875	(70)	700	(55)
Shrubs (leaves and twigs)	22	140	(9)	140	(9)	230	(14)	250	(15)	1100	(70)	700	(45)
Conifers (branch tips)	20	880	(30)	600	(20)	100	(3)	100	(3)	4100	(145)	2000	(70)
Average, all plants	67	480	(27)	200	(11)	150	(8)	150	(8)	2000	(115)	1000	(55)
Soils													
		Mean		Median		Mean		Median		Mean		Median	
"B" horizon	39	<80		<200		45		30		660		500	

*Analyses for zinc–deciduous trees, 9 spectrographic and chemical analyses; grass, 15 chemical analyses; shrubs, 9 chemical analyses; conifers, 18 chemical analyses; and average, all plants, 42 chemical analyses.

50 ppm at 50 to 100 m (≈165–325 ft). They suggested that mosses be used as indicators of pollution because mosses near highways accumulate lead to a remarkable degree, with contents of as much as 700 ppm in dry weight. They concluded from their calculations of lead quantities that "only a minor part of the lead, which is liberated by the cars on the combustion of local petrol, will settle and accumulate in the vicinity of the roads." This conclusion bears out that made by Tatsumoto and Patterson (1963) that the concentrations of lead they found in the surface water of the oceans could be attributed to gasoline. Tatsumoto and Patterson have also found lead with an isotopic composition similar to that of gasoline lead in the snows of the Sierras. Chow (1970) has shown by isotopic analysis that petrol lead occurs in surface soil, but not at depth.

Table 2 shows the mean content for lead in the ash and computed in the dry weight of 106 vegetables and 58 soils, which were collected from gardens at various distances from roads in the southeastern part of Washington County, Maryland. A significant increase due to contamination produced by vehicular traffic can be seen in the lead content of produce that grows within 25 ft of a road. In all collections, however, the median values were within the expected (low) ranges; the mean values were elevated by sporadically high lead values. These occasional high values probably can be attributed to contamination from lead-arsenate sprays used in particular gardens; corn samples from fields where spraying would not be expected were consistently low in lead. Other factors that influence the data are the part of the plant sampled and the preparation of the sample designed to conform with that customary in the processing for human consumption. Stems, for instance, contain more lead than leaves, and leafy vegetables more than fleshy fruits and vegetables, whereas peeled vegetables are consistently lower in lead than those that are unpeeled. Maximum contents of 500 ppm measured in the ash, or 55 to 85 ppm computed in the dry weight, were found in cabbage and unpeeled tomatoes grown within 10 ft of a road. The ratio of lead in vegetation to that in the soil is much higher where the source of lead is from cars than where the source is geologic, suggesting that exhaust lead is absorbed directly from the atmo-

TABLE 2. MEAN LEAD CONTENTS, IN PARTS PER MILLION, OF GARDEN VEGETABLES AND SOILS AS RELATED TO DISTANCE FROM ROADS IN WASHINGTON COUNTY, MARYLAND

	Vegetables		Soils
Feet from road	Ash	Dry weight	Dry weight
1–25	47.0 (41)*	5.7 (41)	253 (13)
26–100	12.9 (26)	1.1 (26)	112 (13)
100–1000	16.2 (20)	2.1 (20)	132 (18)
1000–5000	13.7 (19)	0.93 (19)	70 (14)

*Number in parentheses is the number of samples analyzed. Analyst: E. F. Cooley

sphere. Spanish moss (*Tillandsia usneoides*), which lives entirely on nutrients collected from the atmosphere, has been sampled near major highways in Georgia. A maximum of 1.5-percent lead has been found in the ash and 585 ppm in the dry weight (Shacklette, 1969, written commun.).

Some data on lead in grass along highways in the Denver, Colorado, area were published in 1962 (Cannon and Bowles, 1962) from collections made in 1961. Grass samples along one of the four Denver traverses (south from 6th Avenue) were again collected at the same stations in 1969. The contents of 9 metals in the latter collection are given in Table 3. Both washed and unwashed grass from this collection have been analyzed for lead by atomic absorption (a) on dry plant material using wet digestion (dried at no greater than 50°C) and (b) on plant ash. The plant ash also has been analyzed by emission spectrograph. The lead contents of the unwashed grass were not appreciably higher than those of the washed grass, but the collections were made following a series of heavy rains. No significant difference was found in lead content of grass that had been submitted to drying and ashing before analysis and the lead in grass that had been extracted by wet digestion, after the values in dry-weight percentages were converted to percentages in the ash. Also, the analyst has been unable to extract any measurable lead from the dried plant sample by using organic solvents.

Extreme differences in lead and zinc contents were found, on the other hand, in grass collected at the same station after an 8-yr lapse in time. The average lead content of grass along the 1000-ft traverse has risen nearly 1000 percent in the 8-yr interval, and that of zinc, 260 percent, as shown in Table 4. Increased levels of lead, zinc, iron, and also cadmium extend over the entire traverse. Nickel, strontium, and magnesium contents in the grass are higher within 100 ft of the highway. Such a large increase in metal contamination accompanying increased vehicular traffic is cause for alarm. Contaminated unwashed pasture grass has a greater potential for affecting the health of grazing animals than prepared vegetables have for affecting human health. In addition to the levels of metal reported in Tables 3 and 4, some metal in free dust is presumed to be ingested, along with the grass, by stock pastured in these areas, particularly during periods of drought.

Rühling and Tyler (1968) grew plants from seed in contaminated soils obtained near highways. The plants contained only 5 to 10 ppm lead in dry weight of the plant, whereas weeds that grew in the same soils along the roadsides contained 68 to 950 ppm lead in the dry weight of washed samples. These data suggest direct sorption of lead from the atmosphere by the leaves. Our data suggest, however, that the lead absorbed is contained in the leaves in a relatively stable and probably inorganic form.

Mining and Treatment Plants

Smelters and refineries, commonly located in well-populated areas, may contribute to the atmosphere considerable amounts of the particular metal being refined. Rare byproduct metals are also released to the atmosphere

TABLE 3. METAL CONTENT, IN PARTS PER MILLION, IN WASHED PASTURE GRASS COLLECTED IN 1969 IN RELATION TO DISTANCE FROM A MAJOR HIGHWAY, DENVER, COLORADO

Element	Near Highway	Distance from highway (ft)							
		5	10	50	100	200	500	750	1000
Lead:									
Ash	1,500	3,000	1,500	700	700	700	700	700	300
Dry wt.	105.	222.	109.	56.	56.7	56.7	59.5	50.7	27.7
Nickel:									
Ash	15	13	20	15	15	10	10	10	10
Dry wt.	1.0	0.96	1.4	1.2	1.2	0.81	0.85	0.72	0.92
Zinc:									
Ash	580	400	360	300	280	280	300	280	300
Dry wt.	40.6	29.6	26.1	24.	22.6	22.7	25.5	20.3	27.7
Strontium:									
Ash	700	700	500	1000	1000	300	300	300	300
Dry wt.	49.	51.8	36.	80.	81.	24.3	25.5	21.7	27.7
Boron:									
Ash	200	100	200	150	100	150	200	100	50
Dry wt.	14.	7.4	14.5	12.	8.1	12.1	17.	7.2	4.6
Chromium:									
Ash	15	15	15	15	20	7	10	15	5
Dry wt.	1.0	1.1	1.1	1.2	1.6	.57	0.85	1.1	.46
Iron									
Ash	10,000	7,000	7,000	10,000	15,000	7,000	7,000	10,000	7,000
Dry wt.	700.	518.	507.	800.	1215.	567.	595.	725.	647.
Cadmium:									
Ash	1.8	1.8	1.6	1.2	1.4	1.2	1.6	1.9	1.3
Dry wt.	.13	.13	.12	.096	.11	.1	.14	.14	.12
Magnesium:									
Ash	15,000	15,000	10,000	10,000	10,000	7,000	7,000	7,000	7,000
Dry wt.	..	..	..	..	..	..	..	..	..

Analysts: B. W. Lanthorn, Thelma F. Harms
Note: No decrease noted in titanium, manganese, barium, molybdenum, copper, vanadium, or zirconium values.

TABLE 4. A COMPARISON OF LEAD AND ZINC CONTENTS, IN PARTS PER MILLION, OF GRASS COLLECTED IN 1961 AND 1969 FROM THE SAME STATIONS NEAR DENVER, COLORADO

Feet from highway	Lead in June 8, 1961 collection		Lead in May 12, 1969 collection		Lead percentage increase from 1961 to 1969	Zinc in June 8, 1961 collection		Zinc in May 12, 1969 collection		Zinc percentage increase from 1961 to 1969
	In Ash	(Dry wt.)	In Ash	(Dry wt.)		In Ash	(Dry wt.)	In Ash	(Dry wt.)	
2 (scraped)	500	(38.)	1500	(105.)	200	350	(25.2)	580	(40.6)	65.7
5	315	(28.)	3000	(222.)	850	100	(9.0)	400	(29.6)	300.
10	145	(12.)	1500	(109.)	934	100	(7.8)	360	(26.1)	260.
50	95	(8.1)	700	(56.)	636	50	(4.4)	300	(24.0)	500.
100	125	(11.2)	700	(57.)	460	100	(10.0)	280	(22.6)	180.
200	50	(4.0)	700	(57.)	1300	50	(3.9)	280	(22.7)	460.
500	35	(2.8)	700	(60.)	1900	100	(8.3)	300	(25.5)	200.
750	25	(2.2)	700	(51.)	2700	50	(4.4)	280	(20.3)	460.
1000	25	(1.9)	300	(28.)	1100	50	(3.9)	300	(27.7)	500.

Analysts: E. F. Cooley, B. W. Lanthorn, C. S. Papp

from the stacks of smelters and refineries if preventive measures are not taken. The amount of any one element, as shown in Table 5, depends upon the type of ore and its geologic source and, of course, the method of processing.

Contamination of Vegetation. Commonly, vegetation is killed for a considerable distance around active smelters. Canney (1959) described an area, 3 mi in diameter, around the smelter near Kellogg, Idaho, as completely denuded. The barren region is surrounded by a border zone of grass, stunted shrubs, and small trees. A similar picture can be observed in most smelter and refinery environs.

The contamination of vegetation by *arsenic* and *copper*, near the Anaconda, Montana, smelter, was noticed in the early 1900s by Swaine and Harkins (1908). They reported as much as 1800 ppm copper in hay, but claimed that a large part of this was copper adherent to the surface of the leaf; hairy plantain leaves contained the most copper, and the bark of trees was high in copper as compared to the wood. Warren and others (1966) found as much as 22,000 ppm *zinc* in the ash of poplar leaves and 1600 ppm *lead* in the ash of birch leaves 3 mi from the Trail smelter in British Columbia.

Cadmium is readily absorbed by vegetation and appears to concentrate in the woody parts of the plants (Table 6). Branches from a pinyon collected recently 0.6 mi from the old Leadville, Colorado, smelter site (inoperative for the past 13 years) contained 18,000 ppm lead in the ash (342 ppm by dry weight) and 140 ppm cadmium in the ash (2.7 ppm by dry weight). The needles contained considerably less of both elements. Sedge collected along drainage 1000 ft from the smelter contained less lead, but contained 20,000 ppm zinc in the ash or 1960 ppm calculated in the dry weight.

Collections were also made near the smelter at Bartlesville, Oklahoma, where ores from the lead-zinc deposits of Missouri are treated (Table 6). Lead values were considerably lower, but cadmium values were, in general, higher than those obtained near the Leadville smelter. Sedge appears to concentrate zinc (20,000 ppm), but not cadmium. High values of 11,200 ppm zinc and 1025 ppm cadmium were found in the ash of bark of poplar growing in soil containing 3.2 percent zinc at a distance of 1500 ft from the Bartlesville smelter. Edible onions growing at a distance of 7000 ft from the smelter contained as much as 230 ppm cadmium in the ash or 12.6 ppm calculated in the dry weight.

The Bartlesville plant samples were analyzed by atomic absorption; measurements were made on the ash in one series, and on the dry plant material in another. No significant loss in ashing was detected.

Mills in a mining district are also a source of contamination for trees. Junipers, which normally contain 1 ppm *uranium* in the ash, contained as much as 1000 ppm near the Uravan mill in Colorado, an average of 150 ppm at a distance of 800 to 1500 ft, and an average of 40 ppm at 2000 to 4000 ft (Cannon, 1952).

Contamination of Soils. Pollution of soils near treatment plants also has been studied. According to H. E. Hawkes (1957, written commun.) soils within 1 mi of the *copper* smelter at Superior, Arizona, average 0.5 percent (5000 ppm) copper at the surface. Soils near the smelter at Bartlesville, Oklahoma (Table 6), contained, in a populated area, more than 3 percent (30,000 ppm) zinc, 0.25 percent (2500 ppm) lead, and 0.045 percent (4500 ppm) cadmium. Canney (1959) has shown that the soil within 1000 ft of the smelter near Kellogg, Idaho, in the Coeur d'Alene district, contains about 0.25 percent (2500 ppm) *zinc* and a similar amount of *lead*. The degree of contamination is sufficient to prevent the use of soil sampling as a means of prospecting for at least 5 mi from the smelter, unless samples are collected at a depth of more than 10 in. Garland Gott (*in* Gott and others, 1969; 1970, oral commun.) reported that a belt about 1 mi wide extending away from the smelter on the windward side is contaminated with *cadmium* and *antimony* that are derived from the smelter. He reported that cadmium, which occurs in residual soil in mineralized areas in a ratio of 1 cadmium to 120 zinc, was emitted at one time from the stack of the smelter in a ratio of 1 cadmium to 1 zinc. Thus, he has been able to outline the distribution of smelter contamination from cadmium by plotting ratios of zinc to cadmium at soil sample stations in the district.

This increment of atmospheric metal contaminating both soils and plants should be an important consideration in regulating the local pasturing of stock and the raising of vegetables in mining districts and areas of industrial and urban pollution.

Mineral Deposits in Place

Volatile elements or compounds may be released directly to the air from ore deposits in place. This phenomenon has been proved for *mercury,* which is associated with many base- and precious-metal deposits. Mercury-in-air detectors have, therefore, been developed for the detection of mercury as a pathfinder element in prospecting for heavy-metal deposits (McCarthy and others, 1969). By this method, mercury in soil gas and in the atmosphere is collected by amalgamation on gold or silver foil and is subsequently released and measured by an atomic-absorption instrument. Initial airborne tests show that levels remain high for several hundred feet above ground and are much increased during warm weather as shown in Figure 1. Mercury in air at 200 ft above base-metal deposits measures as much as 20 times the established background of 4.5 nanograms per cubic meter; some measurements are given in Table 7. Preliminary results show that the depth of overburden does not have an adverse effect on the amount of mercury released to the atmosphere and, also, that there is a diurnal variation in mercury-in-air contents, which is inversely proportional to the barometric pressure.

TABLE 5. CONTENT, IN PERCENT, OF RARE METALS IN FLUE DUST FROM REFINERIES AND SMELTERS, UNITED STATES

Refinery or smelter and type of flue dust	Year collected	Antimony	Bismuth	Cadmium	Cobalt	Indium
Silver refinery, Golden Cycle mill, Cripple Creek district, Colorado ("dust from base of stack")	1942	L*	L	L	0.004	L
Hegeler Zinc Co. smelter, Danville, Illinois, custom concentrates ("flue dust, grab sample")	1942	L	L	0.35	0.004	0.008
White Knob Copper Co. smelter, Mackay, Idaho ("flue dust from dump")	1942	L	L	L	L	L
Washoe Reduction Works, Anaconda, Montana ("copper flue dust, grab sample")	1945	0.14	0.036	.044	L	L
U.S. Steel Co. Provo steel plant, Provo, Utah ("flue dust")	1943	L	L	L	.008	L
Eastern United States ("copper flue dust from blast furnaces")	1943	.007	L	.0009	.008	L
Eastern United States ("dry flue dust from iron roasters")	1943	.001	L	L	.008	L

TABLE 5. (CONTINUED)

Refinery or smelter and type of flue dust	Year collected	Molybdenum	Nickel	Selenium	Silver	Tin	Vanadium
Silver refinery, Golden Cycle mill, Cripple Creek district, Colorado ("dust from base of stack")	1942	0.03	0.004	0.032	. .	L	0.028
Hegeler Zinc Co. smelter, Danville, Illinois, custom concentrates ("flue dust, grab sample")	1942	.0007	.008	. .	. .	0.0036	.0045
White Knob Copper Co. smelter, Mackay, Idaho ("flue dust from dump")	1942	.007	.0008	.01	. .	.0009	.0028
Washoe Reduction Works, Anaconda, Montana ("copper flue dust, grab sample")	1945	.001	L	. .	0.005	.009	.002
U.S. Steel Co. Provo steel plant, Provo Utah ("flue dust")	1943	.007	.02	.0008	. .	L	.17
Eastern United States ("copper flue dust from blast furnaces")	1943	.004	.004	. .	. .	.017	L
Eastern United States ("dry flue dust from iron roasters	1943	.005	.006	. .	. .	.0009	.0006

*L, less than 0.0005 percent; leaders (. .), not detected. Samples were analyzed spectrographically by J. C. Rabbitt, except those for Se which were analyzed chemically by H. W. Lakin. Data from Kaiser and others (1954).

TABLE 6. ZINC, LEAD, AND CADMIUM CONTENTS OF PLANTS AND SOILS, IN PARTS PER MILLION, AS RELATED TO SMELTER CONTAMINATION IN OKLAHOMA AND COLORADO

				Zinc		Cadmium		Lead		
		Distance in ft	Part of plant	Ash	Converted to dry wt.	Ash	Converted to dry wt.	Ash	Converted to dry wt.	Run on dry plant
	Bartlesville, Oklahoma									
D413–387	*Populus deltoides*	1,500	Leaves	9,200	(856)	450	(42.0)	120	(11.2)	15.0
–410	do.	do.	Wood	5,100	(66)	820	(11.)	190	(2.5)	<2.5
–411	do.	do.	Bark	11,200	(762)	1,025	(70.)	500	(34.)	30.
–406	Top 1/2 in. soil	do.		22,000		170		2,500		
–407	1 to 5 in. soil			32,000		450		2,500		
D412–390	*Ulmus americana*	7,000	Leaves	3,750	(225)	96	(5.8)	120	(7.2)	10.
–388	*Asparagus officinalis*	do.	Tops	1,200	(72)	96	(5.8)	40	(2.4)	5.
–389	*Allium porrum*	do.	Bulb	9,200	(206)	230	(12.6)	200	(11.)	15.
–408	Humus			9,000		85		1,000		
–409	Soil			12,000		7		125		
	Leadville, Colorado*									
D413–452	*Carix* sp.	1,000	Tops	20,000	(1,960)	55	(5.4)	500	(49.)	. .
–453	*Pinus edulis*	3,200	Needles	5,000	(80)	30	(.48)	2,500	(40.)	. .
–454	do.	do.	Wood (incl. bark)	5,200	(99)	140	(2.7)	18,000	(342.)	. .

Analysts: Thelma F. Harms and Clara S. E. Papp.
*Smelter inoperative since 1956

WATER POLLUTION

Large quantities of metals find their way into the rivers and are transported in solution and as suspended matter to the sea. The medians and ranges for 16 of the 18 trace metals in 15 major North American rivers, as computed by Durum and Haffty (1963), are given in Table 8. They show that aluminum, barium, iron, manganese, strontium, and zinc have the greatest range in values, partly because rivers that drain different parts of the country have a characteristic chemical composition. For example, rivers that drain glaciated parts of the continent contained more titanium, chromium, nickel, and manganese; west-coast drainage was higher in molybdenum and lead; and rivers that drain alkaline basins from Texas to Nevada contained unusual amounts of fluorine, strontium, lithium, and boron.

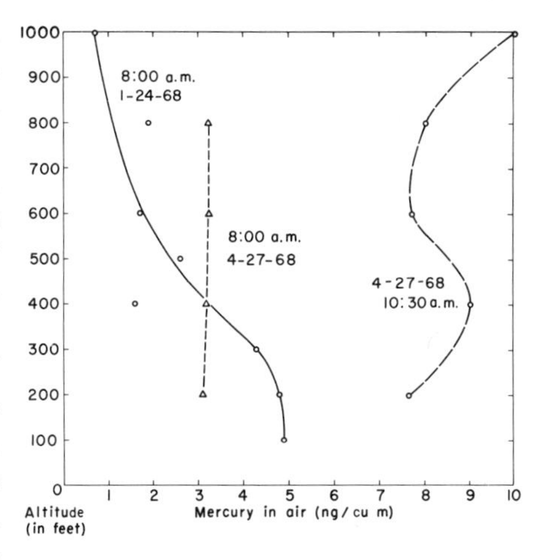

Figure 1. Mercury in air as a function of altitude. From McCarthy and others (1969), and J. H. McCarthy (1969, written commun.).

TABLE 7. MERCURY CONTENT OF AIR, IN NANOGRAMS PER CUBIC METER, AT 200-FOOT ELEVATION OVER ORE DEPOSITS IN ARIZONA*

Area	Minimum	Maximum	Mean
Measurements taken over mercury concentrations			
Superstition Mountains (occurrence)†	58.0	66.0	62.0
Dome Rock Mountains (ore deposit)	12.	57.5	31.4
Measurements taken over porphyry copper deposits			
Ajo	12.	30	18.8
Silver Bell	18.5	53.2	27.6
Measurements taken, for comparison, over barren ground (background concentrations)			
Along highway from Quartzite, Arizona, to Colorado River at Blythe, California	1.6	7.2	4.5

*Collection made by means of aircraft. From McCarthy and others (1969).

†The term "occurrence" is generally used to denote a small area of mineralized ground as opposed to an ore deposit of economic value. The anomaly encountered in the Superstition Mountains was not associated with any known mineralization and may indicate an undiscovered deposit of economic value.

TABLE 8. METAL CONTENT, IN MICROGRAMS PER LITER, OF 15 MAJOR NORTH AMERICAN RIVERS*

	Median	Range	Limits of detection (in 1 liter of water containing 100-mg dissolved solids)
Aluminum	238.0	12 -2550	0.3
Barium	45	9 - 152	.1
Boron	10	1.4 - 58	1
Cobalt	0	0 - 5.8	1
Iron	300	31 -1670	.3
Lead	4	0 - 55	1
Lithium	1.1	.075- 37	.1
Manganese	20	0 - 185	.5
Molybdenum	.35	0 - 6.9	.3
Nickel	10	0 - 71	.5
Rubidium	1.5	0 - 8	.4
Silver	.09	0 - .94	.05
Strontium	60	6.3 - 802	.1
Titanium	8.6	0 - 107	1
Vanadium	0	0 - 6.7	.5
Zinc	0	0 - 215	50

*Rivers are: Apalachicola, Atchafalaya, Colorado, Columbia, Hudson, Mississippi, Mobile, Sacramento, Susquehanna, Yukon, Churchill, Fraser, MacKenzie, Nelson, and St. Lawrence (Durum and Haffty, 1961).

In the eastern states, underground coal-mine workings and old waste piles add acidity, total *iron, manganese,* and *sulfate* to stream drainage, particularly after flush-outs by storm runoff (Corbett and Agnew, 1968). Industrial pollution is also high. A good example is the situation observed on Long Island where large areas of ground water in two counties were contaminated with *chromium* by aircraft plants during World War II (Lieber and Welsch, 1954) and remained polluted for many years. Water analyses by a more sophisticated method in 1954 revealed a similar pattern of contamination from cadmium for 3200 ft south of the plant, reaching a maximum of 3.2 ppm and affecting 22 wells. At a water consumption rate of 1 liter a day, 1.5 to 3.0 mg cadmium would be ingested by an individual per day, an amount reported by Lieber and Welsch to be nearly toxic. The ingestion of local produce, which also has been contaminated, could add additional cadmium to the already high daily intake.

Working metal mills of the west are now required under the Water Quality Acts of 1965 and 1966 to remove the calcium, magnesium, and iron that have been added in the extraction process, and also all radioactive materials, from the effluents before disposal.

Nevertheless, large quantities of metal remain in the tailings collected before 1965 and commonly are in contact with surface drainage. Analyses of the tailings from the Durango, Colorado, uranium mill, of native vegetation growing on alluvium 1 mi downstream, and of vegetables 23 mi downstream from the mill show the following relationships (Table 9).

TABLE 9. METAL CONTENT, IN PARTS PER MILLION, OF MILL TAILINGS AT DURANGO, COLORADO, COMPARED WITH THAT OF VEGETATION DOWNSTREAM

Element	Tailings	1 mi below mill	23 mi below mill	
		Native vegetation (3)* (ash)	Vegetables (55) (ash)	Average of worldwide vegetation (320–3,000) (ash)
Barium	500	175	≈150	1,100
Boron	70	165	175	580
Cobalt	1,000	N	<5	7
Copper	1,000	170	50	165
Manganese	5,000	220	100	3,300
Molybdenum	10	<10	<7	13
Lead	300	70	<10	41
Vanadium	>10,000	<15	<10	20
Zinc	10,000	1,170	≈400	1,220
Percent of ash	. .	6.4	12.2	10.1

Note: Contents of Fe, Ti, Cr, Mn, Sr, and Zr were the same or less than the world average.
*Figures in parentheses represent number of samples averaged. N, not detected.
Analysts: M. D. DeValliere, Thelma Harms, and B. W. Lanthorn.

TABLE 10. METAL CONCENTRATIONS AS RELATED TO CONTAMINATION OF TREES IN LEAD-ZINC DISTRICTS OF MISSOURI

		Zinc (ppm)		Cadmium (ppm)		Lead (ppm)	
		In ash	Converted to dry weight	In ash	Converted to dry weight	In ash	Converted to dry weight
Poplar on abandoned mine dumps at Oronogo, Missouri							
D413–360	Leaves	7,200	684	410	40	120	11
383	do.*	10,000	900	520	47	110	10
361	Twigs	3,280	370	400	45	400	45
362	Wood	4,000	192	460	22	350	17
363	Bark	4,300	344	460	37	300	24
396 + 397	Dump Material	13,700	. .	18	. .	5,000	. .
Yellow birch along drainage from workings north of Joplin, Missouri							
D413–364	Leaves	18,800	1,220	280	18	340	22
384	do.*	18,400	1,435	220	18	350	27
365	Twigs	32,400	1,230	370	14	4,600	175
366	Wood	28,000	420	425	6.5	900	14
367	Bark	27,200	680	840	21	2,900	73
368	Roots	25,200	882	430	15	1,600	50
400 + 401	Humus	21,000	. .	125	. .	2,000	. .
398 + 3399	Alluvium	28,500	. .	160	. .	2,000	. .

*Duplicate sample
Analysts: Thelma F. Harms and Clara S. E. Papp

Vegetables grown in this geographic area are normally very low in metals including those that are produced by the mill. Although nine metals show a considerably higher content in the tailings than in an average soil, only lead and zinc show a slight increase in the native vegetation that grows on alluvium within a mile of the mill downstream, and this questionably higher content has dissipated within the 23-mi distance to the next town.

Trees that grow directly in mineralized soil, particularly in the poorly drained, disturbed areas of mining districts, are subject to severe contamination, as shown in Table 10. Trees growing around old diggings at Joplin, Missouri, contained nearly 2 percent (20,000 ppm) *zinc*, a half percent (5000 ppm) *lead*, and a tenth of a percent (1000 ppm) *cadmium* in the ash of the twigs and bark of yellow birch. The values shown are considerably higher for cadmium than have previously been reported in the literature. The expansion of mining towns that are situated in geographically favorable areas for growth should be planned carefully in order to avoid the building of suburban subdivisions on leveled dump areas; those areas could possibly prove hazardous to health through contamination of garden vegetables.

A final word should be said regarding a common every-day source of pollution. Water used for domestic purposes or for irrigation can be contaminated by the pipes and fittings through which it flows. A comparison was made of the zinc content of water taken directly from the well and from the house faucet at several homes in the high *zinc* area of Manning, New York. The differences are:

Farm	Aquifer type of rock	Depth (feet)	Zinc content in water taken directly from well (ppm)	Zinc content in water taken from house faucet (ppm)
Gaylord	Dolomite	73	8.3	10.0
Shepherd	Till	33	.2	1.0
Dickerson	do.	18	.16 (high H_2S)	6.0

The recent changeover to *copper* pipes in the United States will lower the daily intake of zinc considerably and at the same time raise that of copper. In some geographic areas, this increase may be enough to create nutritional imbalance by changing the ratio of copper to molybdenum. Optimum standards for a given metal content in domestic water thus should vary geographically depending upon the concentrations of other naturally occurring metals in the water. The addition of many metals, termed "pollutants" under present definition, may be beneficial in certain geochemical provinces.

SUMMARY

The geochemist can contribute vitally important information with respect to the pollution problem by providing data on the average amounts of metals in rocks, soils, plants, and water of the natural environment for comparison with contaminated areas. As an example, median values of seven metals in four classes of vegetation and in soils collected in remote areas more than 1000 ft from a road in unmineralized areas have been calculated and presented in Table 1. These samples were collected in various geologic environments that were representative of the conterminous United States, whereas plants and soils from a particular geologically definable area may contain amounts of trace metals that are considerably above or below these averages. It is important for those working on pollution problems to have this information also, as areas in which excesses of lead, cadmium, or other metals occur naturally, may be particularly vulnerable to health problems resulting from additional pollution. The geochemist can also monitor changes in this natural environment caused by various types of air and water pollution from inorganic substances. Increments of metals that contaminate the air near urban areas and, hence, the natural environment result largely from coal burned for industrial uses and from vehicular gasoline. The content of lead and other metals in vegetation growing along major highways is increasing at alarming rates. Lead in grass at stations resampled after a period of 8 yrs has increased nearly 1000 percent.

Areas of mining and smelting are a source of considerable contamination in a radius of several thousand feet. In addition to the major metals of the ore being smelted, anomalous amounts of minor metals, such as selenium, bismuth, cadmium, vanadium, antimony, or arsenic, may be released to the atmosphere. Several thousand parts per million of metals in the ash of local vegetation may be absorbed from smelter effluent. Volatile elements also may be released to the atmosphere from ore deposits in place. Mercury, which is associated with many types of ore deposits, is commonly 20 times background for a height of 200 ft above ground.

Water pollution of a geochemical nature in the western part of the United States is commonly related to coal mining, which adds acidity, iron, manganese, and sulfate to surface waters. Western mining of complex ores releases major ore metals to drainage systems, but the effects on soils and vegetation downstream are commonly slight. In areas of metal deficiency, small additions to surface and ground waters may actually be beneficial. On the other hand, contamination of soils, ground water, and vegetation around mines and treatment plants may be hazardous because minor metals, such as cadmium, bismuth, antimony, and arsenic, that remain in the waste are markedly increased relative to the original percentages in the ore. Tree ash samples containing as much as 3-percent (30,000-ppm) zinc, 0.46-percent (4600-ppm) lead, and 0.1-percent (1000-ppm) cadmium have been collected in populated areas near mining or smelting operations.

The geochemical data presented in this report show the levels of inorganic contamination in air, water, soils, and plants that should be expected from natural and man-controlled sources, and point out the hazards of pollution from metal-mining operations, particularly in areas of naturally high metal content.

City planning in areas known to be mineralized should take into consideration the high metal content of mine and smelter waste dumps and the increments of atmospheric metal that may have been absorbed by soils and vegetation over a considerable area. More information of the type illustrated should be accumulated and made available to scientists involved in the pollution problem to provide them with a sound basis for pollution control.

REFERENCES CITED

Canney, F. C., 1959, Geochemical study of soil contamination in the Coeur d'Alene district, Shoshone County, Idaho: Mining Eng., v. 214, p. 205–210.

Cannon, H. L., 1952, The effect of uranium-vanadium deposits on the vegetation of the Colorado Plateau: Am. Jour. Sci., v. 250, no. 10, p. 735–770.

——, 1955, Geochemical relations of zinc-bearing peat to the Lockport dolomite, Orleans County, New York: U.S. Geol. Survey Bull. 1000-D, p. 119–185.

Cannon, H. L., and Bowles, J. M., 1962, Contamination of vegetation by tetraethyl lead: Science, v. 137, no. 3532, p. 765–766.

Chow, T. J., 1970, Lead accumulation in roadside soil and grass: Nature, v. 225, p. 295–296.

Corbett, D. M., and Agnew, A. F., 1968, Coal mining effect on Busseron Creek Watershed, Sullivan County, Indiana: Indiana Univ. Water Res. Research Center Rept. Inv., no. 2, 187 p.

Dunn, J. T., and Bloxam, H. C. L., 1932, The presence of lead in the herbage and soil of lands adjoining coke ovens and the illness and poisoning of stock fed thereon: Soc. Chem. Industry Jour., v. 51, p. 100–102.

——, 1933, Occurrence of lead, copper, zinc, and arsenic compounds in atmospheric dusts, and the sources of these impurities: Soc. Chem. Industry Jour., v. 52, p. 189–192.

Durum, W. H., and Haffty, Joseph, 1961, Occurrence of minor elements in water: U.S. Geol. Survey Circ. 445, 11 p.

——, 1963, Implications of the minor element content of some major streams of the world: Geochim. et Cosmochim. Acta, v. 27, no. 1, p. 1–11.

Gott, G. B., Botbol, J. M., Billings, J. M., and Pierce, A. P., 1969, Geochemical abundance and distribution of nine metals in rocks and soils of the Coeur d'Alene district, Shoshone Country, Idaho: U.S. Geol. Survey open-file Rept., 3 p.

Kaiser, E. P., Herring, B. F., and Rabbitt, J. C., 1954, Minor elements in some rocks, ores, and mill and smelter products: U.S. Geol. Survey TEI-415, 119 p. (rept. prepared *for* U.S. Atomic Energy Comm.).

Lieber, Maxim, and Welsch, W. F., 1954, Contamination of ground water by cadmium: Am. Water Works Assoc. Jour., v. 46, no. 6, p. 541–547.

McCarthy, J. H., Jr., Vaughn, W. W., Learned, R. E., Meuschke, J. L., 1969, Mercury in soil gas, and air—A potential tool in mineral exploration: U.S. Geol. Survey Circ. 609, 16 p.

Rühling, Åke, and Tyler, Germund, 1968, An ecological approach to the lead problem: Botaniska Notiser., v. 121, p. 321–342.

Shacklette, H. T., Hamilton, J. C., Boerngen, J. G., and Bowles, J. M., 1971, Elemental composition of surficial materials in the conterminous United States: U.S. Geol. Survey Prof. Paper 574-D.

Singer, M. V., and Hanson, Lowell, 1969, Lead accumulation in soils near highways in the Twin Cities metropolitan area: Soil Sci. Soc. America Proc., v. 33, no. 1, p. 152–153.

Swaine, D. J., and Harkins, W. D., 1908, Arsenic in vegetation exposed to smelter smoke: Am. Chem. Soc. Jour., v. 30, p. 915–928.

Tatsumoto, Mitsunobu, and Patterson, C. C., 1963, Concentrations of common lead in some Atlantic and Mediterranean waters and in snow: Nature, v. 199, no. 4891, p. 350–352.

Trelease, S. F., and Beath, O. A., 1949, Selenium; its geological occurrence and its biological effects in relation to botany, chemistry, agriculture, nutrition, and medicine: published *by* the authors, 292 p.

Warren, H. V., and Delavault, R. E., 1960, Observations on the biogeochemistry of lead in Canada: Royal Soc. Canada Trans., 3d ser., v. 54, p. 11–20.

Warren, H. V., Delavault, R. E., and Cross, C. H., 1966, Mineral contamination in soil and vegetation and its possible relation to public health, *in* Pollution and our environment: Canadian Council Resource Ministers Natl. Conf., Montreal, Background Paper A3-3, 11 p.

PUBLICATION AUTHORIZED BY THE DIRECTOR, U.S. GEOLOGICAL SURVEY, WASHINGTON, D.C.

MANUSCRIPT RECEIVED BY THE SOCIETY APRIL 10, 1970

Printed in U.S.A.

THE GEOLOGICAL SOCIETY OF AMERICA, INC.
MEMOIR 123, 1971

Trace Elements Related to Cardiovascular Disease

H. MITCHELL PERRY, JR., M.D.
Washington University School of Medicine
St. Louis, Missouri

ABSTRACT

Two possible relationships between cardiovascular disease and trace elements have been suggested: (1) Studies on human patients and on laboratory animals reveal what may be a causal role for cadmium in hypertension. (2) Comparison of death rates and hardness of drinking water have revealed a statistically significant negative correlation between hardness of water and death rates due to cardiovascular diseases.

CONTENTS

INTRODUCTION

Two major areas where trace elements may relate to cardiovascular disease seem worthy of discussion. It should be emphasized at the outset that no clearcut proof of any such relationships exists, and that no mechanism for them has been postulated. Having made that point let me begin by briefly summarizing the two parts of my presentation.

A relationship between cadmium and hypertension has been suspected on the basis of the following: (1) Certain drugs used in the treatment of human hypertension are strong metal-binding agents. (2) In rats, the chelating agent EDTA can lower elevated blood pressure to normal. (3) Patients with very severe hypertension have increased urinary cadmium. (4) During childhood and adolescence, man accumulates cadmium in his kidneys where it is bound to a peculiar protein of unknown function. (5) The concentration of cadmium in human kidneys varies from one part of the world to another. (6) In the United States, hypertensive patients have more renal cadmium than do normotensive controls. (7) In rats, injected cadmium induces acute hypertension, while chronically fed cadmium induces gradual hypertension. (8) The renal cadmium concentration of cadmium-fed hypertensive rats approximates that of adult Americans.

The second area where trace elements have been suspected of relating to cardiovascular disease involves some constituent of hard water and arteriosclerosis. In both the United States and Great Britain, death rates from cardiovascular disease have been inversely correlated with hardness of water. Separate figures for 1950 and for 1960 have shown this relationship, and the correlation is highly significant, statistically. No other environmental or social factors could be found to explain the correlations. Examining various constituents of hard water revealed no constituent that gave better correlations than calcium carbonate, the major one.

Both areas of possible interaction between trace elements and cardiovascular disease are still highly speculative. At the same time, however, both warrant further multidisciplinary investigation with the reasonable possibility that additional information may aid in the prevention or treatment of major human diseases. On the one hand, involvement of cadmium in human hypertension has certainly not been proved, but it seems reasonable to seek biologic effects from the considerable amounts of cadmium that accumulate in human kidneys and to study the mechanism by which cadmium raises blood pressure in rats. On the other hand, the association between arteriosclerosis and soft water may well be real, but as yet nothing is known regarding the mechanism of the effect, and it is here that further work should be directed.

PART 1. THE POSSIBLE RELATIONSHIP BETWEEN HYPERTENSION AND CADMIUM

The story which culminated in the suspicion that cadmium was related to human hypertension (Schroeder, 1965) began in the early 1950s, soon after effective antihypertensive drugs first became available. It was quickly recognized that ability to bind transition and related trace metals was a common characteristic of several of these drugs that seemed to have a direct relaxing effect on blood vessels. The drugs in question included substituted hydrazines

(Gross and others, 1950), nitroprusside (Page and others, 1955), azide (Black and others, 1954), thiocyanate (Olsen, 1950), and some experimentally effective mercaptans (Schroeder and others, 1955a).

On the basis of this observation, it was decided to test whether other metal-binding agents affected blood pressure. Ethylenediamine tetra-acetate (EDTA) seemed an ideal compound to study; it was a powerful chelating agent, and its antihypertensive potency had not been examined. Moreover, it apparently had no pharmacologic effects beyond its ability to bind metal. Finally, it had the added advantage of not being metabolized; hence, there was no problem of possible pharmacologic effects from its breakdown products. As is shown in Figure 1, ethylenediaminetetraacetic acid (EDTA) rapidly lowered the diastolic pressure of a hypertensive, but not of a normotensive rat. The pressure of the hypertensive animal was lowered to normal, but not below normal levels. In addition, the effect was both temporary and reversible; the pressure returned to its original high level after several hours. It should be emphasized that the neutral calcium disodium salt of EDTA was used in order to avoid binding sodium, potassium, magnesium, or calcium. Thus, the only change in the concentration of any of these four metals came from whatever release of calcium occurred when it was displaced from the administered EDTA by some more tightly bound tissue metal (Schroeder and Perry, 1955b).

Since EDTA presumably lowered blood pressure by binding some metal for which it had a greater affinity than it had for calcium, an attempt was made to approximate its binding strength for this unknown metal by treating hypertensive animals with a series of increasingly tightly bound EDTA chelates of divalent transition metals. The results of these experiments were compatible with copper, zinc, or cadmium as the metal in question (Schroeder and Perry, 1955b).

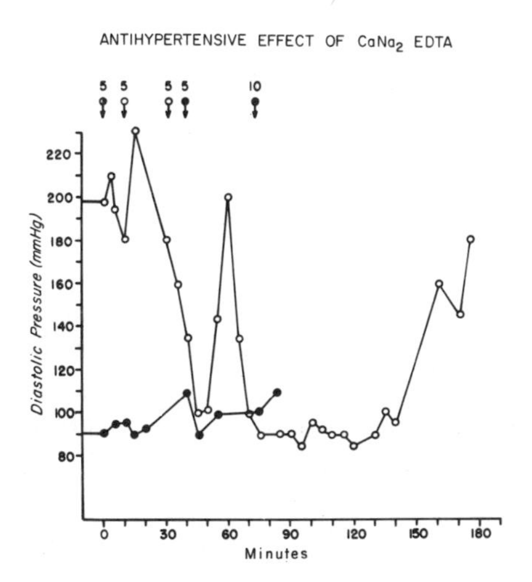

Figure 1. Effects on two typical anesthetized 200 gm rats: open circles relate to a rat with renovascular hypertension which received a total of 15 mg, and solid circles to a normotensive rat which received 20 mg of intravenously injected neutral $CaNa_2EDTA$. Reproduced by permission of The Journal of Laboratory and Clinical Medicine.

The obvious next step was to determine whether EDTA had any effect on human hypertension. The answer was that it had no marked or consistent effect, although in some individuals there was some fall in pressure. One possible explanation for the difference in its antihypertensive effectiveness in human patients was a difference in the etiology of the hypertension. The animal hypertension, which had responded so well, was renovascular—

the only kind that could be consistently produced at the time. The human hypertension, on the other hand, was probably "essential" rather than renovascular, although techniques for differentiating the two were then inadequate (Schroeder and Perry, 1956).

EDTA, however, was not without effect in man. It uniformly did two things: It lowered circulating cholesterol, and it greatly increased urinary zinc. Specifically, 3 gm daily of intravenously injected $CaNa_2EDTA$ produced an immediate ten-fold increase in renal excretion of zinc, and this persisted with relatively little change for a full week of chelate administration. The fall in circulating cholestrol was more gradual and required four days to reach a minimum value which was about three quarters of the control value. The lowered level persisted as long as the EDTA was continued; however, both urinary zinc and blood cholesterol had returned to their control levels within 48 to 72 hours after the last dose of EDTA (Perry and Camel, 1960).

EDTA changed the urinary concentrations of some other trace metals, but not of all of them. Manganese behaved like zinc, with an immediate and persistent five- to ten-fold increase in excretion that disappeared as soon as the EDTA was discontinued. Cadmium and lead also increased at least ten-fold, but their patterns of excretion were somewhat different. For cadmium, the increase was delayed, and it persisted for several days after the last EDTA. For lead, the increase was immediate, but the level returned to nearly normal before the EDTA was discontinued, presumably because the available lead was exhausted. Urinary vanadium and nickel were not affected by EDTA (Perry, 1961).

The next step in seeking a relationship between trace metals and human hypertension was to compare the trace-metal content of urine from normal individuals and hypertensive patients. In addition to a slight, seemingly generalized increase in the urinary excretion of many metals by hypertensive patients, there was a major difference in urinary cadmium concentrations that averaged less than 1 μg per liter for the normotensive subjects as compared with almost 50 μg per liter for the hypertensive patients (Perry and Schroeder, 1955). It must be emphasized that most of the hypertensive patients studied had severe disease with marked renal involvement; hence, it is possible that metal that was bound to renal protein appeared in the urine as kidney tissue was destroyed. Control of the blood pressure tended to lower the excretion of most trace metals by hypertensive patients. In particular, the cadmium concentration fell from its pretreatment level of nearly 50 μg per liter to approximately 5 μg per liter (Perry and Schroeder, 1955; Schroeder and Perry, 1956).

The foregoing observations raised the question as to whether cadmium might be involved in human hypertension. At the time this question arose, the unique distribution of cadmium in the human body with its very marked predilection for the kidney was just being emphasized (Tipton and others, 1953). All adult Americans were discovered to have relatively large amounts in their kidneys, the average individual having a total of more than 10 mg of renal cadmium (Tipton and Cook, 1963). Little renal cadmium was found at birth, the concentration in the newborn being less than 1 percent of the adult concentration (Perry and others, 1961). Renal cadmium was noted to accumulate throughout most of life, but particularly during the first two decades (Schroeder and Balassa, 1961). The usual concentration of cadmium in the adult kidney was observed to be 10 to 20 times that in the liver, and the concentration in the liver was usually several times that of any other organ (Perry and others, 1962).

Concentrations of renal cadmium were found to vary with the geographic origin of the individual, as is shown in Figure 2. Considering subjects from nine different countries, the median values for the concentration of renal cadmium ranged from 720 μg per gm of ash for Burundians to more than seven times that much for Japanese. At the 1 percent level of significance, Caucasoid patients from the United States, Switzerland, and India differed from Negroid Burundians, on the one hand, and from four groups of Mongoloid Asiatics, on the other, having more renal cadmium than the former and less than the latter (Perry and others, 1961).

Finally, hypertensive Americans were reported to have significantly more renal cadmium than similar normotensive subjects. One-hundred and seventeen patients without hypertension who died sudden accidental deaths had an average of 2940 μg of cadmium per gm of kidney ash; 17 hypertensive patients who died similar deaths had 4220 μg per gm of ash. The difference is highly significant; $p<0.0005$ (Schroeder, 1965). An investigator seeking corroboration of this finding, however, reported no significant difference in renal or hepatic cadmium concentration between normal individuals and those with hypertensive cardiovascular disease (Morgan, 1969).

In 1960, purification and characterization of the cadmium-binding protein from equine kidneys were described (Kägi and Vallee, 1960, 1961). This protein which they named metallothionein, had a molecular weight of approximately 10,000. Just over 8 percent of its weight was cadmium and zinc. The total number of atoms of these two metals per molecule of metallothionein was constant, but their ratio was variable. Metallothionein proved to be a unique protein in that about a quarter of its amino acid residues were cysteine, the sulfhydryl groups of which were responsible for the metal binding. More recently, metallothionein has been found in human kidney and liver, apparently in sufficient amounts to account for all of the cadmium in these organs (Pulido and others, 1966).

Our animal studies on the effects of cadmium on blood pressure began more than 10 years ago. We started by chronically feeding rats small doses of cadmium in an attempt to produce hypertension. Weanling animals were fed Purina rat chow to which cadmium had been added to make the final concentration 25 parts per million. Striking hypertension appeared in almost all of the cadmium-treated animals within three months; control animals without cadmium remained normotensive. These results have never been published because subsequent confirmatory experiments were inconclusive, perhaps

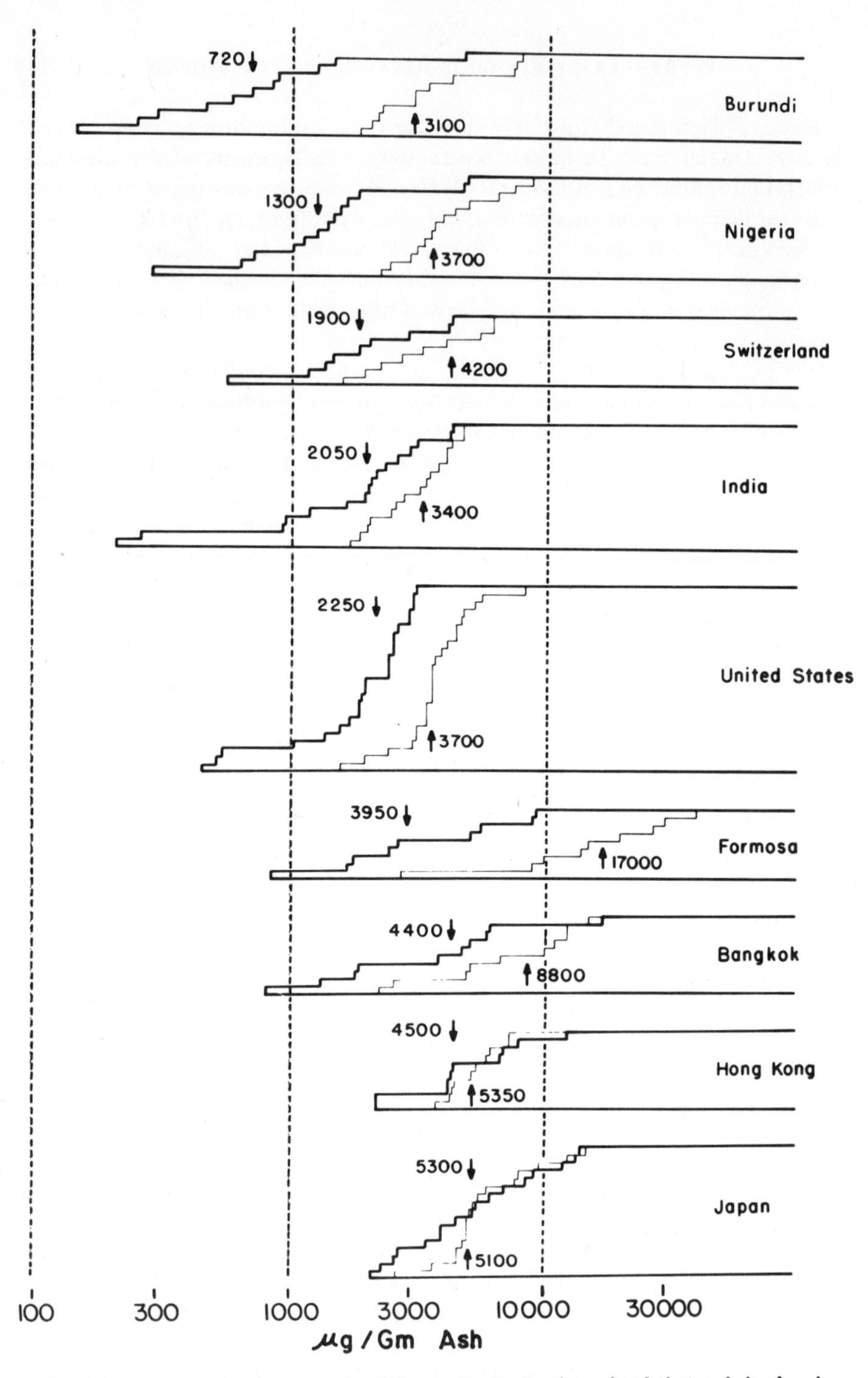

Figure 2. Cumulative frequency graphs of the renal concentrations of cadmium and zinc for nine national groups comprising 119 adult subjects: Data for cadmium are shown by heavy lines and data for zinc by finer lines. Figures cite and arrows point to median values. The height of each of the nine graphs indicates the number of patients from the country in question, with each vertical segment representing one subject. For instance, in the top graph each of the 11 subjects from Burundi is represented by one step in the graph. The horizontal scale is logarithmic. Reproduced by permission of The Journal of Chronic Diseases.

because of methodologic difficulties. It was not until Schroeder developed his special animal care facilities that a similar experiment was successfully repeated (Schroeder and Vinton, 1962). He put the cadmium in drinking water instead of in food; he used a lower dose of cadmium, 5 parts per million instead of 25; and he followed his rats for much longer periods of time, 30 months instead of 3 months. At 17 months he found that half of his cadmium-fed animals were hypertensive, that is, had systolic pressures more than 35 mm mercury above the mean value for rats of comparable age. Using the same criterion, no control animals were hypertensive at that age. At 30 months, two-thirds of the cadmium-fed animals had become hypertensive, although by that time one-eighth of the untreated rats had also become hypertensive. The difference between the two groups, however, remained highly statistically significant. This experiment is only one of several in which Schroeder has induced hypertension by feeding cadmium (Schroeder, 1964; Schroeder and others, 1965).

Although the significance of this hypertension will remain uncertain until more is known about its mechanism, survival among Schroeder's rats given 5 parts per million of cadmium from the time of weaning was less than survival among control animals; the difference was statistically significant (Schroeder and others, 1965). Schroeder has subsequently shown that his cadmium-fed rats had renal cadmium concentrations approximating those of normal American adults (Schroeder and others, 1968). He has recently come to relate the presence of hypertension to renal cadmium-zinc ratios, rather than to absolute values of cadmium. Moreover, he has reported reversal of cadmium-induced hypertension by feeding a zinc chelate which presumably bound cadmium and released zinc, thus depleting renal cadmium and repleting renal zinc (Schroeder and Buckman, 1967).

The mechanism of the very slowly appearing chronic hypertension induced by cadmium-feeding has proved difficult to study; however, in exploring the physiologic effects of cadmium, it was observed that small amounts of the metal given intra-arterially were acutely hypertensive. From 10 to 40 μg of cadmium per 100 gm of rat weight had striking pressor effects, raising the diastolic pressure an average of nearly 30 mm mercury, but only for a few minutes. The response to a second similar dose of cadmium, given after the pressure had returned to normal, was essentially the same as the response to the first dose. Relatively large doses of either EDTA or hydralazine were able to reverse completely the acute hypertension induced by cadmium and to prevent the expected increase in diastolic pressure following a second similar dose of cadmium. Larger doses of cadmium were vasodepressor, and all concentrations of zinc were without effect on blood pressure (Perry and Yunice, 1965).

Further investigation of parenterally administered cadmium revealed: (1) The hypertension induced by intraarterial cadmium was related to increased cardiac output (Perry and others, 1967). (2) Although intravenous cadmium was less effective than intraarterial cadmium in inducing hypertension (Perry and others, 1970), intraperitoneal cadmium induced an equally marked and much more protracted hypertension (Perry and Erlanger, in press). (3) There is a marked and prolonged increase in renin activity in the peripheral blood following intraperitoneal cadmium and a lesser increase following oral cadmium (Perry and Erlanger, 1970).

The story must be left at this point, incomplete and of uncertain significance. Hypertension has been reported in animals following long-term ingestion of cadmium. Moreover, an acute *in vivo* pressor effect of parenterally administered cadmium has been clearly demonstrated and studies on the mechanism of this effect are in progress. The significance of these cadmium-induced increases in blood pressure and their relationship, if any, to "essential" human hypertension will probably not be known until we have much more information about the mechanism of the animal hypertension and information on the biological effects of the accumulated renal cadmium in man.

PART 2. THE POSSIBLE RELATIONSHIP BETWEEN ARTERIOSCLEROSIS AND HARDNESS OF WATER

The first serious suggestion that cardiovascular diseases might be related to the hardness of drinking water was made by Schroeder using data from the United States (Schroeder, 1960a). He pointed out that, in 1950, average state-wide, age-adjusted annual death rates[1] from all causes varied greatly. They ranged from a maximum of 983 per 100,000 in South Carolina, a "soft water" state, to a minimum of 712 in Nebraska, a "hard-water" state. He further observed that percentage differences from state to state were even larger for the death rates due to cardiovascular disease, which varied from 511 to 290. Using the reports of the National Office of Vital Statistics (1949–1951) on the one hand and the analyses of the United States Geological Survey (1950–1951) on the other, Schroeder sought correlations between death rates and weighted average hardness of water.[1] To his surprise he found that total death rates were inversely related to hardness of water. The correlation coefficient, calculated on a state by state basis, was −0.36, which was statistically significant, with $p<0.01$.[1]

Total deaths in the United States can be divided into two approximately equal groups: cardiovascular[1] and all other. The correlation coefficient for state-wide death rates from cardiovascular diseases and average hardness of

[1]To minimize repetition of cumbersome phraseology, the following simplifications will be used: (1) "Death rate(s)" will stand for average age-adjusted annual rate(s) per 100,000 population. (2) "Hardness of water" will stand for a weighted average hardness obtained by weighting the hardness of each supply by the size of the population using it; hardness is the total calcium and magnesium in parts per million. (3) Designation of correlations as "suggestive," "significant," or "highly significant" will indicate "p" values of <0.05, <0.01, or <0.001, respectively. (4) "Cardiovascular" will refer to categories 330–334 and 400–468 of the "International List of Diseases and Causes of Death;" moreover, parenthetical numbers following the designation of a disease entity will identify it according to that list.

water was −0.56; for death rates from all other causes and hardness of water, it was +0.02 (Schroeder, 1960a).

Systematically subdividing cardiovascular disease and considering a high-risk subpopulation (white, 55 to 64 years), Schroeder found that, on a state by state basis, hardness of water was significantly correlated with death rates from hypertension with heart disease (440–443) and arteriosclerotic heart disease including coronary disease (420). Also related, but less strongly so, were death rates from vascular diseases of the central nervous system (330–334), general arteriosclerosis (450), and hypertension without mention of heart (444–447). The correlation coefficients and their significances, as calculated by Schroeder, are cited in Table 1 (Schroeder, 1960b).

Making similar calculations on the basis of 163 large metropolitan areas in the United States instead of on a statewide basis, Schroeder found a similar strong relationship between hardness of water and coronary death rates of middle-aged white males (Schroeder, 1960a). In an effort to pinpoint the causal relationship, he also examined the coronary death rates in the 163 metropolitan areas with respect to 16 constituents of their water. Concentrations of calcium, magnesium, bicarbonate, sulfate, and total dissolved solids were found to be as significantly correlated as total water hardness, but no more so (Schroeder, 1960a).

Schroeder also investigated the over-all situation by seeking correlations between hardness of water and statewide death rates for each of the ten leading causes of death. For the entire population of the United States, he observed, in addition to the previously reported significant negative correlation for cardiovascular disease, one suggestive and two significant correlations: (1) a suggestive negative correlation for vascular lesions affecting the central nervous system (330–334), the third most frequent cause of death; (2) a significant positive correlation for motor vehicle accidents (E810–835), the fourth most frequent cause of death; and (3) a significant positive correlation for congenital malformations (750–759), the ninth most frequent cause of death. Schroeder speculated that the relationship with motor vehicle accidents was probably a "geographic accident" resulting from laxer speed laws in the sparsely populated hard-water states of the midwest; he did not interpret the relationship with congenital malformations. Schroeder considered the obvious question about the effect on cardiovascular disease of the "geographic accident" of hardwater states being less densely populated than the soft water states; he felt that the correlation between hardness of water and cardiovascular death rate could not be explained by a correlation between that rate and urban living (Schroeder, 1960b).

In 1961, the year after Schroeder's two reports from the United States, Morris and co-workers reported a similar negative correlation between cardiovascular death rates and hardness of water in Great Britain. Considering white males from 45 through 64 years of age, these investigators found that in 83 county boroughs of England and Wales, death rates from cerebrovascular disease (330–334), arteriosclerotic heart disease (420), and myocardial degen-

TABLE 1. CORRELATIONS IN THE UNITED STATES DURING 1950 BETWEEN HARDNESS OF WATER AND DEATH RATE FROM VARIOUS CARDIOVASCULAR CONDITIONS

Cause of Death	Annual U.S. Death Rate Per 100,000	Correlation Coefficient (r)	Statistical Significance (p)
All Causes	844	−0.36	0.01
All noncardiovascular	420	+0.02	NS
All cardiovascular (330–334, 400–468)*	424	−0.56	<0.01
Hypertension without mention of heart** (444–447)	12	−0.29	<0.05
Hypertension with heart disease** (440–443)	100	−0.57	<0.01
General arteriosclerosis** (450)	11	−0.34	<0.02
Vascular lesions affecting central nervous system** (330–334)	170	−0.34	<0.02
Arteriosclerotic heart disease** (420)			
Male	817	−0.50	<0.01
Female	267	−0.45	<0.01

Data taken from Schroeder (1960b). Figures in the first three lines represent all ages and races and both sexes; remaining figures are for high risk subpopulations. NS indicates not significant.

*Parenthetical figures indicate categories from "International List of Diseases and Causes of Death."

**Data for subcategories of cardiovascular diseases pertain to white persons, 55 to 64 yrs of age.

eration (422) had highly significant correlations with hardness of water. Hypertension with heart disease (440–443) had a suggestive correlation; whereas hypertension without heart disease (444–447) was unrelated. Bronchitis was the only noncardiovascular disease found to be suggestively related to hardness of water, and it, too, was negatively correlated. Table 2 presents a comparison between these data and the 1950 American data (Morris and others, 1961).

Morris and co-workers also sought correlations between water hardness and various "environmental indices," including income, education, social class, population density, cigarette smoking, domestic overcrowding, motor vehicle accidents, and industrial air pollution. They found no meaningful correlations except for a suggestive one with industrial air pollution. When their correlations were recalculated to eliminate the suggestive effect of pollution, the correlation between water hardness and death rates from bronchitis dropped sharply, while that between water hardness and cardiovascular death rates was scarcely affected. These investigators then sought correlations between cardiovascular death rates and various components of water. They found highly significant correlations with total hardness, carbonate and calcium; they found no correlations with magnesium or sulfate (Morris and others, 1961).

The next year Morris and co-workers made several additional observations. They reported that lead, zinc, iron, and copper were not present in excess in any of the waters examined and, hence, were probably not related to the variations in cardiovascular death rates associated with water hardness. They observed that both rainfall and latitude were positively correlated with cardiovascular death rates, that is, the greater the rainfall or the higher the latitude, the larger the death rate. The well-known relationship between rainfall and water hardness would explain the former but not the latter, since latitude and rainfall were not related (Morris and others, 1962).

Subsequently, Schroeder examined 1960 data to determine whether the correlation between total cardiovascular death rates and water hardness which he had reported for 1950 was again present. This correlation and those for both hypertensive and coronary disease were still significant, while the correlation for cerebrovascular disease was still only suggestive (Table 2; Schroeder, 1966). Crawford, Gardner, and Morris examined 1961 data in search of the correlations they had observed for 1951 data. They too found the correlations they had previously reported; moreover, their new correlations were even stronger than their earlier ones (Table 2). Again, social factors could not be correlated significantly with cardiovascular death rates, and this time climatic factors were only suggestively correlated (Crawford and others, 1968).

Recently, Masironi has restudied the relationship of cardiovascular death rates to the hardness of local water supplies, using several different approaches. For one of these, he chose four American rivers with a considerable range of hardness and compared the death rates for people living in the river basins.

Using alpha-radioactivity, a water quality measurement which he found significantly correlated with hardness, Masironi found that, while non-cardiovascular death rates were virtually the same along the Ohio, Columbia, Missouri and Colorado Rivers, death rates from arteriosclerotic heart disease and hypertensive heart disease decreased as alpha-radioactivity increased, being respectively 25 percent and 41 percent lower along the Colorado (which had the hardest water) than along the Ohio (which had the softest water). To control the factor of urbanization, he compared the data after matching the counties along the same river basins for population, and again found death rates from arteriosclerotic heart disease and hypertensive heart disease to be lower for the hard water, high alpha-radioactivity rivers (Masironi, in press).

In addition to these five essentially consistent sets of data, several less extensive investigations should be listed. As early as 1957, Kobayashi suggested that hard water was related to a decrease in apoplectic deaths in Japan (Kobayashi, 1957). A few years later Muss noted an inverse relation between hardness of water and cardiovascular death rates for middle-aged white American males (Muss, 1962), and in Sweden, Biörck and coworkers found hardness of water to be inversely related to cerebrovascular disease, coronary disease, and myocardial degeneration (Biörck and others, 1965). In Holland, Biersteker (1967), and in Sweden,Böstrom and Wester (1967) found that hard water was related to decreased cardiovascular mortality among women, but not men. No such relationship could be found in Oklahoma by Lindeman and Assenzo (1964) or in Ireland by Mulcahy (1964). A small population studied in England showed that the calcium content of water was inversely related to cardiovascular death rates (Robertson, 1968), while another in Czechoslovakia showed the opposite (Ošancova and Hejda, 1968). Anderson in Ontario found water hardness inversely correlated only with the rates of sudden deaths from ischemic heart disease and not with the non-sudden ones (Anderson and others, 1969), an observation supported by Peterson (1970) in America.

Although the basis of the relationship between hard water and cardiovascular disease remains unknown, Crawford and Crawford have attempted to go beyond mere correlations and to gain some insight into the mechanisms which might be involved. Specifically, they sought to determine whether there was more anatomic arteriosclerosis in the soft water areas or whether anatomically similar hearts were more susceptible to a given insult in those areas. To this end they compared hearts from Glasgow, a very soft-water area, with hearts from London, a very hard-water area. Death rates from coronary disease were known to be much higher in Glasgow than in London, being 613 and 398 per 100,000, respectively, for men from 45 through 64 years old. These workers considered two series of subjects: 121 men who died sudden accidental deaths and 196 men who died of coronary disease. Among the young subjects without clinical coronary disease, the coronary arteries of the Scots had more "confluent atheromata" and more "lumen stenosis" than did those of the Englishmen. Moreover, at all ages the Scots had considerably

Table 2. Correlations Between Hardness of Water and Death Rate from Various Cardiovascular Conditions in Two American and Two British Studies

	United States		Great Britain	
	1949–1951	1960	1951 Series	1961 Series
All Causes	−0.38*	−0.64**	−0.39**	−0.55**
Cardiovascular disease (330–334 and 400–468)	−0.53*	—	−0.54**	−0.65**
Cerebrovascular disease (330–334)	(−0.34)	(−0.25)	−0.42**	−0.56**
Arteriosclerotic heart disease (420)	−0.48*	(−0.51)**	−0.39**	−0.52**
Myocardial degeneration (422)	—	—	−0.47**	—
Hypertension with heart disease (440–443)	(−0.57)*	(−0.42)**	—	—
Hypertension without heart disease (444–447)	(−0.29)	—	—	—
Noncardiovascular disease	−0.10	—	−0.19	−0.35*

The values cited are correlation coefficients "r". The data are taken from Schroeder's two American series of patients (Schroeder 1960a, 1960b, 1966) and from Morris, Crawford, and co-workers' two British series of patients (Morris and others, 1961, Crawford and others, 1968; see text). The American data are for white subjects; the British data, although actually for all races, can also be considered as being for white subjects. Values without parentheses apply to men from the ages of 45 through 64. Parenthetical values for the first American series refer to all patients from 55 through 64 years of age. Parenthetical values for the second American series refer to all males. Blanks in the table represent unavailable data.

Asterisks indicate the statistical significance of the cited values. No asterisk indicates that p is greater than 0.01. For values with 1 or 2 asterisks, p is respectively less than 0.01 or 0.001. In the first American series, p's were simply given as less than 0.01, although some "r's" suggest higher levels of significance.

After each "Cause of Death," its numerical designation from the "International List of Disease and Causes of Death" is given parenthetically.

more myocardial fibrosis and healed infarcts, suggesting that the environment at Glasgow produced more anatomic changes. The increased anatomic changes in the Scots, however, were not reflected in increased vascular accumulations of alkaline earths, at least not among the younger subjects, since the Englishmen under 40 years of age had an average of 20 percent more calcium and 50 percent more magnesium than the Scots in their coronary arteries (Crawford and Crawford, 1967).

Among the young subjects with fatal coronary disease, the Scots had more old coronary obstructions, healed myocardial infarcts, and recent infarcts than the Englishmen. Thus, anatomic changes of arteriosclerosis were more frequent in Glasgow, both in patients with and in patients without clinically evident coronary disease. On the other hand, limited anatomic disease, as judged by extent of atheromata and degree of stenosis, was present in more Scots with fatal coronary disease than in the comparable Englishmen—one-third of the former, but only one-sixth of the latter—suggesting that for some patients at least soft water might be associated with an increased tendency toward fatal myocardial infarction over and above any tendency toward increased severity of arteriosclerotic lesions (Crawford and Crawford, 1967). Parenthetically, in 1951 Morris had suggested that the incidence of myocardial infarction depended on factors other than anatomic change. This suggestion stemmed from his observation that for a London hospital population there were more infarcts, but less anatomic change in the late 1940s than there had been at the turn of the century (Morris, 1951). Finally, the Crawfords found that among their patients with fatal coronary disease, more Scots than Englishmen had low concentrations of calcium and magnesium in their coronary arteries, so that both metals were apparently accumulated in larger amounts in hard-water areas (Crawford and Crawford, 1967).

In closing, two additional epidemiologic studies are worthy of mention. The first emphasizes the potential danger of drawing conclusions from mere correlations without benefit of supporting theory since it demonstrates that coronary mortality correlates equally well with parameters other than hardness of water. The second suggests that in some tenuous and as yet undefined way, the element cadmium may serve to link the two otherwise apparently separate parts of my presentation. In the first, Friedman reported that coronary death rates were correlated with per capita cigarette sales. Using average statewide data for the 33 states which licensed cigarette sales in 1950, the correlation coefficient was 0.76 for middle-aged white males. Friedman also found coronary death rates correlated with water hardness. Moreover, partial correlations revealed that the significant effect of water hardness remained after the effect of cigarette consumption had been eliminated. In other words, his data indicated that cigarettes could not account for the previously demonstrated correlation (Friedman, 1967).

In the second study, Carroll reported that for 28 American cities death rates from all types of heart disease (420–447), except rheumatic, were significantly correlated with the cadmium content of urban air, with the correlation coefficient "r" being 0.76. He also found that coronary (420) and hypertensive (440–447) death rates were similarly correlated, with "r" being 0.67 and 0.61, respectively. In contrast, cerebrovascular disease was not significantly correlated; "r" was negative. The association could not be explained by high prevalence of both polluted air and heart disease in urban areas, since there was no correlation with other indices of air pollution such as suspended particulate matter and benzene-soluble organic materials. In fact, zinc was the only other air pollutant significantly correlated with heart disease (Carroll, 1966).

In summary, then, the currently available data suggest that death rates for the entire group of cardiovascular diseases may be related to hardness of drinking water, although the explanation for this remains obscure. Coronary disease seems to be more generally involved than any other single diagnostic entity, since there was a significant negative correlation between its death rate and water hardness in both American and both British series. In the two American series, hypertension with heart disease also had a high negative correlation. In the two British series, cerebrovascular disease had a high negative correlation.

REFERENCES CITED

Anderson, T. W., le Riche, W. H., and MacKay, J. S., 1969, Sudden death and ischemic heart disease; correlation with hardness of water supply: New Eng. J. Med., v. 280, p. 805.

Biersteker, K., 1967, Hardness of drinking water and mortality: Tijdschr. Soc. Geneeck, v. 45, p. 658.

Biörck, G., Boström, H., and Wistrom, A., 1965, On the relationship between water hardness and death rate in cardiovascular diseases: Acta Med. Scandinavia, v. 178, p. 239.

Black, M. J., Zweifach, B. W., and Speer, F. D., 1954, Comparison of hypotensive action of sodium azide in normotensive hypertensive patients: Soc. Exper. Biology and Medicine, v. 85, p. 11.

Böstrom, H., and Wester, P. O., 1967, Trace elements in drinking water and death rate in cardiovascular disease: Acta Med. Scand., v. 181, p. 465.

Carroll, R. E., 1966, The relationship of cadmium in the air to cardiovascular disease death rates: Am. Med. Assoc. Jour., v. 198, p. 177.

Crawford, T., and Crawford, M. D., 1967, Prevalence and pathological changes of ischaemic heart-disease in a hard-water and in a soft-water area: Lancet, v. 1, p. 229.

Crawford, M. D., Gardner, M. J., and Morris, J. N., 1968, Mortality and hardness of local water-supplies: Lancet, v. 1, p. 827.

Friedman, G. D., 1967, Cigarette smoking and geographic variation in coronary heart disease mortality in the United States: Jour. Chronic Disease, v. 20, p. 769.

Gross, R., Druey, J., and Meier, R., 1950, Eineneuegruppeblutdrucksenkender Substanzen Von Besonderem Wirkungscharakter: Experientia, v. 6, p. 19.

Kägi, J. H. R., and Vallee, B. L., 1960, Metallothionein: A cadmium- and zinc-containing protein from equine renal cortex: Jour. Biol. Chemistry, v. 235, p. 3460.

——, 1961, A cadmium- and zinc-containing protein from equine renal cortex: Jour. Biol. Chemistry, v. 236, p. 2435.

Kobayashi, J., 1957, Geographical relationship between chemical nature of river water and death-rate from apoplexy: Benchte d. Ohara Inst. f. landivertsch. Biologie, v. 11, p. 12.

Lindeman, R. D., and Assenzo, J. R., 1964, Correlations between water hardness and cardiovascular deaths in Oklahoma counties: Am. Jour. Public Health, v. 54, p. 1071.

Masironi, R., in press, Cardiovascular mortality in relation to radioactivity and hardness of local water supplies.

Morgan, J. M., 1969, Tissue cadmium concentration in man: Arch. Intern. Med., v. 123, p. 405.

Morris, J. N., 1951, Recent history of coronary disease: Lancet, v. 1, p. 69.

Morris, J. N., Crawford, M. D., and Heady, J. A., 1961, Hardness of local water-supplies and mortality from cardiovascular disease: Lancet, v. 1, p. 860.

——, 1962, Hardness of local water-supplies and mortality from cardiovascular disease: Lancet, v. 2, p. 860.

Mulcahy, R., 1964, The influence of water hardness and rainfall on the incidence of cardiovascular and cerebrovascular mortality in Ireland: Jour. Irish Med. Assoc., v. 55, p. 17.

Muss, D. L., 1962, Relationship between water quality and deaths from cardiovascular disease: J. Am. Water Works Assn., v. 54, p. 1371.

Olsen, N. S., 1950, Inhibitory effect of thiocyanate upon oxidation mediated by liver and kidney: Arch. Biochemistry, v. 26, p. 269.

Ošancova, K., and Hejda, S., 1968, In: Den Hartog, et al, *Eds.*: Dietary studies and epidemiology of heart diseases: The Hague, Stichting tot Wetenschappelijka Voorlichting op Voedingsbied.

Page, I. H., Corcoran, A. C., Dustan, H. P., and Koppanyi, T., 1955, Cardiovascular actions of sodium nitroprusside in animals and hypertensive patients: Circulation, v. 11, p. 188.

Perry, H. M., Jr., 1961, Discussion: Chelation therapy in circulatory and sclerosing disease: Fed. Proc., v. 20, p. 254.

Perry, H. M., Jr., and Camel, G. H., 1960, Some effects of $CaNa_2EDTA$ on plasma cholesterol and urinary zinc in man, *in* Seven, M., *Ed.*: Metal-Binding in Medicine: J. B. Lippincott Co., Philadelphia, p. 209–215.

Perry, H. M., Jr., and Erlanger, M. W., 1970, Cadmium-induced increase in peripheral renin activity: Circulation, Oct. Suppl.

Perry, H. M., Jr., and Erlanger, M., in press, Hypertension and tissue metal levels following intraperitoneal cadmium, mercury and zinc: Am. J. Physiol.

Perry, H. M., Jr., and Schroeder, H. A., 1955, Concentration of trace metals in urine of treated and untreated hypertensive patients compared with normal subjects: Lab. and Clin. Medicine Jour., v. 46, p. 936.

Perry, H. M., Jr., and Yunice, A., 1965, Acute pressor effects of intra-arterial cadmium and mercuric ions in anesthetized rats: Soc. Exper. Biology and Medicine Proc., v. 120, p. 805.

Perry, H. M., Jr., Tipton, I. H., Schroeder, H. A., Steiner, R. L., and Cook, M. J., 1961, Variation in the concentration of cadmium in human kidney as a function of age and geographic origin: Jour. Chronic Diseases, v. 14, p. 259.

Perry, H. M., Jr., Tipton, I. H., Schroeder, H. A., and Cook, M. J., 1962, Variability in the metal content of human organs: Lab. and Clin. Medicine Jour., v. 60, p. 245.

Perry, H. M., Jr., Erlanger, M., Yunice, A., and Perry, E. F., 1967, Mechanism of the acute hypertensive effect of intra-arterial cadmium and mercury in anesthetized rats: J. Lab. Clin. Med., v. 70, p. 963.

Perry, H. M., Jr., Erlanger, M., Yunice, A., Schoepfle, E., and Perry, E. F., 1970, Hypertension and tissue metal levels following intravenous cadmium, mercury, and zinc: Am. J. Physiol , v. 219, p. 755.

Peterson, D. R., 1970, Water hardness, arteriosclerotic heart disease, and sudden death: Am. Heart Assoc., CVD Newsletter No. 8, p. 24.

Pulido, P., Kägi, J. H. R., and Vallee, B. L., 1966, Isolation and some properties of human metallothionein: Biochemistry, v. 5, p. 1768.

Robertson, J. S., 1968, Mortality and hardness of water: Lancet, v. 2, p. 348.

Schroeder, H. A., 1960a, Relation between mortality from cardiovascular disease and treated water supplies: Am. Med. Assoc. Jour., v. 172, p. 1902.

——, 1960b, Relations between hardness of water and death rates from certain chronic and degenerative diseases in the United States: Jour. Chronic Diseases, v. 12, p. 586.

——, 1964, Cadmium hypertension in rats: Am. Jour. Physiology, v. 207, p. 62.

——, 1965, Cadmium as a factor in hypertension: Jour. Chronic Diseases, v. 18, p. 647.

——, 1966, Municipal drinking water and cardiovascular death rates: Am. Med. Assoc. Jour., v. 195, p. 81.

Schroeder, H. A., and Balassa, J. J., 1961, Abnormal trace metals in Man: Cadmium: Jour. Chronic Diseases, v. 14, p. 236.

Schroeder, H. A., and Buckman, J. B., 1967, Cadmium hypertension. Its reversal in rats by a zinc chelate: Arch. Environmental Health, v. 14, p. 693.

Schroeder, H. A., and Perry, H. M., Jr., 1955, Antihypertensive effects of metal binding agents: Lab. and Clin. Medicine Jour., v. 46, p. 416.

——, 1956, Essential and abnormal trace metals in cardiovascular diseases: Council on high blood pressure research, Ann. Mtg. Proc., v. 14, p. 71.

Schroeder, H. A., and Vinton, W. H., Jr., 1962, Hypertension induced in rats by small doses of cadmium: Am. Jour. Physiology, v. 202, p. 518.

Schroeder, H. A., Menhard, E. M., and Perry, H. M., Jr., 1955, The antihypertensive properties of some mercaptans and other sulfur-containing compounds: Lab. and Clin. Medicine Jour., v. 45, p. 431.

Schroeder, H. A., Balassa, J. J., and Vinton, W. H., Jr., 1965, Chromium, cadmium, and lead in rats: Jour. Nutrition, v. 86, p. 51.

Schroeder, H. A., Nason, A. P., and Mitchener, M., 1968, Action of a chelate of zinc on trace metals in hypertensive rats: Am. Jour. Physiology, v. 214, p. 796.

Tipton, I. H., and Cook, M. J., 1963, Trace elements in human tissue. Part II. Adult subjects from the United States: Health Physics, v. 9, p. 103.

Tipton, I. H., Foland, W. D., Bobb, F. C., and McCorkle, W. C., 1953, Spectrographic analysis of human tissue: Phys. Review, v. 91, p. 1.

MANUSCRIPT RECEIVED BY THE SOCIETY APRIL 1, 1970

Printed in U.S.A.

THE GEOLOGICAL SOCIETY OF AMERICA, INC.
MEMOIR 123, 1971

Health-Related Function of Chromium

WALTER MERTZ, M.D.
Walter Reed Army Institute of Research
Washington, D.C.

ABSTRACT

Trivalent chromium is essential for maintenance of normal glucose tolerance, growth, and longevity in rats and other laboratory animals. Deficiency of the element results in diminished sensitivity of tissue to insulin, *in vivo* and *in vitro,* suggesting that chromium may act as a co-factor for the hormone. Various pilot studies have demonstrated that marginal chromium deficiency states can exist in man, characterized by chromium-responsive impairment of glucose tolerance.

CONTENTS

HEALTH-RELATED FUNCTION OF CHROMIUM

Chromium is the latest trace element for which a health-related role has been demonstrated. As with every other element, one must realize that there are different phases of action, depending on the administered dose. A trace element may be toxic in very high amounts; it may exhibit a pharmacodynamic action with intermediate doses—but these are not related to the biological action in which we are specifically interested. The latter can be demonstrated only against the background of a deficiency. Our main question is whether deficiencies of chromium exist or can be induced in various populations of man and laboratory animals. Chromium deficiency can be observed in a number of species. In the rat and the squirrel monkey, it is associated with an impaired glucose tolerance (Schwarz and Mertz, 1959; Davidson and Blackwell, 1968). The rate at which an injected glucose load is removed from the blood stream is decreased in chromium-deficient animals to approximately half of values found in normal controls. This impairment, which develops within a few weeks after the low-chromium diet is fed, can be prevented by 2 ppm of the element in diet or drinking water or cured by one stomach-tubed dose of a suitable chromium complex. The impairment of glucose tolerance is the first symptom of a low-chromium state in rats, observable at a time when neither fasting blood glucose levels nor growth rates are abnormal. However, a more severe degree of chromium deficiency resulting in decreased growth and longevity, elevated fasting blood glucose and cholesterol levels and glycosuria, has been produced in animals raised in a strictly controlled environment (Schroeder, 1968).

The mode of action of chromium appears to be closely related to that of insulin, as was first suggested in experiments measuring glucose uptake of epididymal adipose tissue of chromium-deficient rats *in vitro*. Glucose utilization was nearly the same in tissue from deficient and supplemented rats, in the absence of added insulin. However, the tissue from rats raised on a chromium-sufficient diet responded to insulin with a significantly greater increase of glucose uptake than did the chromium-deficient fat pads. The same pattern of response has since been obtained in other systems, *in vitro* and *in vivo*. For example, the *in vitro* addition of chromium significantly enhanced the action of insulin on glucose uptake, oxidation, or conversion to fat. These results prompted a closer investigation of the first insulin-responsive step of glucose metabolism, permeation of sugar from extracellular into intracellular water. The rate of cell entry of D-galactose, a sugar which is not metabolized by peripheral tissue, was slightly increased by the addition of insulin, and the effect of the latter was greatly enhanced by addition of chromium. These results can be interpreted to mean that sugar transport across the cell membrane is one site of action of chromium and that the latter plays a role as a cofactor for the action of insulin. Further evidence for this role can be derived from polarographic studies measuring the reaction between insulin, chromium, and receptor sites on mitochondrial membranes. The results of these studies suggested that chromium participates in a ternary complex, forming between cell-membrane sulfhydryls and the intrachain disulfide of the insulin A chain, facilitating the initial reaction of insulin with its receptors (Mertz, 1967).

If the action of chromium is so closely related with that of insulin, one has to expect that any of the diverse effects of insulin, not only those on glucose metabolism, should be diminished in a chromium-deficient system and should be increased by the addition of trace amounts of chromium. This was shown to be true for three glucose-independent actions of insulin: Stimulation of the transport into heart muscle of α-amino isobutyric acid, a nonutilizable amino acid; increased utilization of 3 amino acids for protein synthesis by heart muscle; and increased swelling *in vitro* of liver mitochondria. In each of these systems the response of the chromium-deficient tissue to insulin could be significantly increased by chromium (Mertz, 1969).

The doses of chromium that produce a significant effect *in vitro* are very small; concentrations of 0.1 to 10 nanograms per 2 ml medium and 100 mg tissue are sufficient. Chromium, like other trace elements, produces a typical dose-response: Activity increases with increasing dose only within a certain range, until a plateau of activity is reached. Increasing the concentration further does not produce additional increases, but often results in decreased function. The effect of oral doses depends on the form in which the element is given. Simple salts of trivalent chromium are poorly absorbed, and amounts of 20 to 50 μg/100 g body weight are required. However, there is evidence suggesting that the element in the form of a natural complex, designated glucose-tolerance factor, is effective in much smaller concentrations. Diets that contain less than 100 ppb are considered chromium deficient; feeding these diets to rats results in the appearance of symptoms. However, chemical analysis by itself is no indicator of the biological availability of the element and must be interpreted with caution.

The effects of chromium deficiency and resupplementation in experimental animals, which have been discussed here, open the question whether low-chromium states also exist in man. There exists an interesting parallelism between the gradual development of impaired glucose tolerance in rats that are fed a low-chromium ration and the well-known declining efficiency of glucose utilization with age in the United States population.

It is difficult to accurately assess the nutritional status of people with regard to chromium. Universally acceptable standards of normal tissue levels do not yet exist, measurements of blood or plasma chromium levels yield little useful information, and no exact knowledge of the daily requirement of man exists. Until such criteria are established, one must rely mainly on indirect evidence.

The first major body of such evidence came from extensive surveys of chromium concentrations in human tissues in the United States and abroad. These studies demonstrated conclusively that the chromium concentrations in most tissues of the United States population decline steadily from birth to old age (Schroeder and others, 1962). This finding is surprising because of the exposure of the large urban population to atmospheric metallic contamination. The latter is reflected in slowly increasing chromium levels in lungs, but these levels are obviously not in equilibrium with the other organ stores. In addition to these age-related differences, there are remarkable regional differences in occurrences and organ concentrations within the United States and between the United States and foreign countries. The average chromium concentration in liver from residents in Dallas, for example, is less than one tenth of the average concentration in livers from residents in the New York or Chicago area.

The average of chromium in liver from foreign countries is more than twice the average United States level. The decline of tissue concentrations with age may be a normal process, but it may also express the development of a deficiency, due to insufficient dietary intake. The fact that higher concentrations are found in the tissues of some foreign populations supports the latter alternative. This would mean that soil and water in some areas contain the element in a form and in quantities that are available to plants, animals and, ultimately, the population. An example of a regional chromium deficiency has recently been established in a study with malnourished children in Jordan (Hopkins and others, 1968). Protein-calorie malnutrition is associated in some areas with impaired glucose tolerance and fasting hypoglycemia. Determination of glucose metabolism in Jordanian refugee children, admitted to a Jerusalem hospital because of malnutrition, revealed that all children from the area around Jerusalem had an impaired glucose tolerance test and fasting hypoglycemia, whereas all children from the Jordan River Valley exhibited a normal test. Both groups were equally malnourished and both came from refugee camps in which the main source of animal protein was a milk powder supplied by the United Nations. The impaired glucose tolerance and fasting hypoglycemia was restored to normal overnight by one oral dose of chromium in all children treated, indicating that chromium deficiency was a complication of the malnutrition state among the children from the Jerusalem area. Analysis of the milk reconstituted from the milk powder detected a chromium content of only 18 ppb, an extremely low concentration. The drinking water from various wells in the Jordan River Valley contained an average of three times as much chromium as the water from various wells and cisterns in the Jerusalem area. It is not certain whether the different chromium intake from drinking water alone (0.5 *versus* 1.5 μg/liter) can account for the different nutritional status with regard to chromium; the higher chromium content that might be expected in the vegetables grown in the valley area may also be an important factor. A study in Nigeria revealed impaired glucose tolerance in malnourished children and improvement following chromium supplementation. Nigerian livers contained only one-third of the chromium concentrations found in the United States, and only one-seventh of the average concentrations in samples outside of the United States. On the other hand, the impaired glucose tolerance of Egyptian children was found not to be related to chromium deficiency, as it did not respond to chromium supplementation and was not associated with low levels of the element (Carter and others, 1968).

Within the population of the United States, assessment of the nutritional chromium status is more difficult. It has been found in several investigations that part of the subjects responded to chromium supplements with an improvement of the previously impaired glucose tolerance (Glinsmann and Mertz, 1966; Levine and others, 1968). In other studies, supplementation was ineffective (Sherman and others, 1968). It must be realized that localized substantial chromium deficiencies are not likely to occur in this country, because of the free flow of nutrients from one area to another. While there are marked differences in the chromium concentration of surface waters within the United States, the municipal drinking waters are remarkably uniform. Yet, there is a portion of the population whose impaired glucose tolerance can be improved with chromium, indicating the existence of a low-chromium state. This may be related to several factors. One of these is the dietary intake of chromium. The average daily intake has been estimated at around 60 μg of the element. Individual dietaries offered in a nursing home or *ad libitum* daily meals of young medical students, measured during one week, provided from 5 to 100 μg of chromium per day (Levine and others, 1968). These were normal and supposedly complete diets. It is obvious that a personal preference for certain foods might result in a low chromium intake that may lead to a depletion of the body stores over a period of time. Most seafoods are low in chromium. Refining of natural sources of carbohydrate is often associated with a substantial loss not only of chromium but also of other elements. This is evident from the lower concentrations in refined sugar than in molasses, in white flour compared to whole grain, or even in white bread compared to whole wheat bread (Schroeder, 1968). Furthermore, consumption of large amounts of sugar leads to a considerable acute rise of the plasma chromium concentration, and much of this increment is subsequently lost in the urine (Schroeder, 1968). But there are also other factors:

Diabetic children have less hair chromium than normal children (Hambidge and others, 1968), and women who have borne children have less than women who have not (Hambidge and Rodgerson, 1969). Some diabetics handle ingested tracer amounts differently than normals. These considerations show that the picture of the nutritional chromium state in different populations is complicated and far from complete. But we know even less when it comes to the very fundamental question of how much chromium is available to plants in the soils of different regions. It is not enough to have available the results of determinations of total chromium in soils; it is necessary to know how much of the total is available to the crop grown in the soil. Studies performed in Germany, France, Poland and Russia have demonstrated that deficiencies do exist in several regions, as shown by very substantial increases of crop yields following trace fertilization with chromium. It appears from the recent work of Bertrand and de Wolf (1968) in France that the chromium extractable from soil with ethylenediaminetetraacetic acid is a good indicator of the available metal. In such soils, the application of 100 g of the element per acre has led to a more than 40 percent increase of crop yields. The advantages of such studies are obvious; Delineation of deficient soils followed by the very inexpensive trace fertilization is not only rewarded by better crop yields but can be expected to contribute ultimately to a better supplementation of the population with chromium.

REFERENCES CITED

Bertrand, D., and de Wolf, A., 1968, Requirement for the trace element chromium in potatoes: C. R. Acad. Sci., v. 266, p. 1494.

Carter, J. P., Kattab, A., Abd-El-Hadi, K., Davis, J. T., El Gholmy, A., and Patwardhan, V. N., 1968, Chromium (III) in hypoglycemia and in impaired glucose utilization in kwashiorkor: Am. Jour. Clin. Nutrition, v. 21, p. 195.

Davidson, I. W. F., and Blackwell, W. L., 1968, Changes in carbohydrate metabolism of squirrel monkeys with chromium dietary supplementation: Soc. Experimental Biol. Med. Proc., v. 127, p. 66.

Glinsmann, W. H., and Mertz, W., 1966, Effect of trivalent chromium on glucose tolerance: Metabolism, v. 15, p. 510.

Hambidge, K. M., and Rodgerson, D. O., 1969, Comparison of hair chromium levels of nulliparous and parous women: Am. Jour. Obstet. Gynecology, v. 103, p. 320.

Hambidge, K. M., Rodgerson, D. O., and O'Brien, D., 1968, Concentration of chromium in the hair of normal children and children with juvenile diabetes mellitus: Diabetes, v. 17, p. 517.

Hopkins, L. L., Jr., Ransome-Kuti, O., and Majaj, A. S., 1968, Improvement of impaired carbohydrate metabolism by chromium (III) in malnourished infants: Am. Jour. Clin. Nutrition, v. 21, p. 203.

Levine, R. A., Streeten, D. H. P., and Doisy, R. J., 1968, Effects of oral chromium supplementation on the glucose tolerance of elderly human subjects: Metabolism, v. 17, p. 114.

Mertz, W., 1967, Biological role of chromium: Fed. Proc., v. 26, p. 186.

——, 1969, Chromium: Occurrence and function in biological systems: Physiol. Rev., v. 49, p. 163.

Schroeder, H. A., 1968, The role of chromium in mammalian nutrition: Am. Jour. Clin. Nutrition, v. 21, p. 230.

Schroeder, H. A., Balassa, J. J., and Tipton, I. H., 1962, Abnormal trace elements in man: Chromium: Jour. Chron. Diseases, v. 15, p. 941.

Schwarz, K., and Mertz, W., 1959, Chromium (III) and the glucose tolerance factor: Arch. Biochemistry Biophysics, v. 85, p. 292.

Sherman, L., Glennon, J. A., Brech, W. J., Klomberg, G. H., and Gordon, E. S., 1968, Failure of trivalent chromium to improve hyperglycemia in diabetes mellitus: Metabolism, v. 17, p. 439.

AUTHOR'S PRESENT ADDRESS: HUMAN NUTRITION RESEARCH DIVISION, AGRICULTURAL RESEARCH SERVICE, BELTSVILLE, MARYLAND 20705

MANUSCRIPT RECEIVED BY THE SOCIETY JANUARY 5, 1970

Printed in U.S.A.

THE GEOLOGICAL SOCIETY OF AMERICA, INC.
MEMOIR 123, 1971

Trace Elements Related to Dental Caries and Other Diseases

FRED L. LOSEE
Eastman Dental Center, Rochester, New York
AND
B. L. ADKINS
University of Queensland, Brisbane, Australia

ABSTRACT

Collaborative studies on the geographic distribution of caries-resistant naval recruits illustrate the type of data available when scientists from many disciplines become interested in a health problem. In a study of the state of Ohio it was found that the residences of 62 percent of the caries-resistant men were on high-lime Wisconsin Till soils associated with rocks that contain extensive celestite deposits. The water supplies had significantly more boron, lithium, molybdenum, and strontium than the finished water supplies for the seven largest cities in Ohio.

A rat-feeding experiment is described in which the water from the caries-resistant area produced a reduction in caries of rats. A vegetable-cooking experiment showed that certain trace elements in the water are taken up by the vegetable and other elements are released to the water from the vegetable. This may be an important factor in daily mineral intake. Selenium concentrations in human blood show an association with geographic differences in selenium distribution in plants. A plea is made for a pooling of resources between environmental geochemists and medical, dental, and animal investigators.

CONTENTS

INTRODUCTION

Three areas selected for study included northwest Ohio, northeast South Carolina, and west-central Florida.

The initial inspection of the regions disclosed that all three were high in agricultural use of the land and that the northwest Ohio area would lend itself best for detailed examination.

It was gratifying to find people at the personal, city, county, state, or federal levels willing to help without compensation, except for the self-satisfaction that comes from contributing to studies that may lead to better health. The Population Branch of the Economic Research Division in the Ohio State Development Department had information on the stability and ethnic breakdown of the population in the area of interest. The Water Division in the Ohio Department of Natural Resources contributed data on the standard water analyses of ground and surface waters within the region. The County Extension Agents for Agriculture and 4-H clubs knew what the soil conditions were and what species and strains of vegetables were grown and used locally. The meat packers could trace the beef and hog from the local area back to the local stores, including chain stores. The same could be done for chickens, turkeys, and dairy products.

In general, we had found that the population was quite a stable one, with many households in the third or older generations. The water used for drinking and cooking, with the exception of two towns, was highly mineralized ground water with a range of fluoride concentrations from 0.2 ppm to 1.8 ppm. The foodstuffs included a very high percentage of locally grown vegetable and animal products, unexpected in this age of rapid transportation and food distribution. Of the 82 caries-resistant recruits from the state of Ohio, 51 men, or 62 percent, came from one soil region, the high-lime Wisconsin Till soils. The soils in this region were developed from glacial till containing considerable limestone and clay. The soil maps show the area as gray-brown podzolic and humic-gley soils developed under forest vegetation. The 1957 Yearbook of Agriculture maps the area as manganese deficient for vegetables, oats, and soybeans. Skougstad and Horr (1963) in a study of the general occurrence of strontium in natural water of the United States, report strontium values uncommonly high for northwest Ohio because of localized celestite deposits ($SrSO_4$).

The uniqueness of the geologic environment in a caries-free area was sufficient to justify more specific examination of the water and foodstuffs.

For the water analyses we were fortunate in obtaining the co-operation of the U.S. Geological Survey. Through the Water Resources Analytical Laboratory at the Federal Center, Denver, Colorado, and the efforts of Marvin Skougstad and Paul Barnett, analyses have been obtained of domestic water used by specific caries-resistant individuals as well as water from the schools they attended. When the median trace-element concentrations of drinking water in the northwest Ohio area are compared to the medians of the finished public water supplies for the seven largest cities in Ohio or the cities in states with a high caries prevalence we found that northwest Ohio has significantly more strontium, 6100 μg/l; molybdenum, 17 μg/l; lithium, 19 μg/l; and boron, 150 μg/l; (Losee and Adkins, 1969). This was of great interest to us because in rat-feeding experiments each of the four elements, at one time or another, has shown some cariostatic or caries-reducing effect when used separately (Büttner, 1963; Shaw and Griffiths, 1961).

EXPERIMENTS

The above findings posed this question: Can the waters from northwest Ohio be shown, experimentally, to produce a reduction in caries experience? Therefore, an exploratory rat-feeding experiment was designed to measure the effect on rat caries of a representative water sample from the northwest Ohio area (Losee and Adkins, 1968). In the diet of experimental rats we used the northwest Ohio water as drinking water and supplemented a cariogenic diet with 1 percent by weight of the ash of green beans cooked in the water. The basic cariogenic diet consists of 63.8-percent sucrose as granulated sugar, 30-percent vitamin-free casein,[1] 2.2-percent Vitamin Diet Fortification Mixture,[1] 2-percent HMW Salt Mixture (Hubbell and others, 1937), and 2-percent corn oil.[1]

A quantity of frozen green beans was purchased in a case lot and was cooked according to the instructions given on the package. Three different waters were used for cooking. The first aliquot of green beans was cooked in distilled water (DW); the next in northwestern Ohio water, which contained 1.5 ppm of fluoride (NW) and which was taken from the kitchen tap of the household of one of the caries-resistant recruits. The third batch was cooked in distilled water which was brought up to 1.5 ppm fluoride by adding NaF (DWF). In each case, the beans were dried at 110°C. and ashed at 450°C.

With the cariogenic diet and DW ash diets, distilled water was supplied for drinking.

With the DWF ash diet, distilled water containing 1.5 ppm fluroide was given, and water from northwest Ohio was used with the NW ash diet.

Before proceeding to describe the outcome of the experiment, we wish to report an important finding in relation to the role of water in cooking. A sample was taken of northwestern Ohio water before cooking commenced and also of the waters which remained after cooking in all three experiments. The four samples were sent for spectrographic analysis to Skougstad and Barnett at the U.S. Geological Survey, and aliquots were sent to M. Little, at Eastman Dental Center, for fluoride analysis. Based on published data, we had anticipated losses of minerals from the green bean to the water in varying quantities, with the exception of fluoride, which has been shown by Smith and others (1945) at the Arizona Agricultural Experimental Station and Martin (1951) at Evanston, Illinois, to be transferred from water to the vegetable.

[1]Nutritional Biochemicals Corporation

Of the elements that could be measured, it was found that aluminum, barium, boron, copper, manganese, nickel, and rubidium were lost from the bean to the water during cooking. In the case of fluoride there was a decrease from 1.5 ppm to 1.1 ppm in the water during cooking. However, there was a striking decrease in the concentration of strontium in the water; it was reduced from 16,000 $\mu g/L$ to 5700, that is, the water contributed 10,300 μg of strontium to each 1000 grams of beans cooked. Molybdenum and lithium also were absorbed by the bean. Thus, of the four elements transferred from the water to the bean, three of them—strontium, molybdenum, and lithium—were among the four elements previously discussed in relation to caries resistance.

Regarding the rat findings, it was obvious that the cariogenic diet, with or without the ash supplement, gave a growth pattern comparable with the commercial laboratory chow for the 28-day experimental period.

In the case of carious lesions, the DW (distilled water) ash and the DWF (with fluorine) ash diets reduce the number, and this reduction is statistically significant with probability <0.05. The NW (Ohio water) ash reduces the number of lesions even further and again the difference between the DW ash or DWF ash on the one hand and NW ash on the other, is statistically significant with probability <0.05 (Losee and Adkins, 1968).

We have a two-stage reduction in caries. In the first stage, we are comparing the prevalence of caries in rats eating the cariogenic diet with the caries experience of those eating the same basic diet, but supplemented with the ash of green beans cooked in either distilled water or distilled water with 1.5 ppm of fluoride. The reduction observed must be attributed to the mineral composition of the ash.

But, to what can we attribute the additional cariostatic effect of the NW ash diet which includes the northwest Ohio drinking water? In this case, we can discount phosphate and fluoride as being the only effective factors, because all three ash-supplemented diets contained the same concentration of phosphorus, and the northwest Ohio water contained no PO_4. In the case of fluoride, the DWF ash diet and water contained the same amount of fluoride as the NW ash diet and NW water.

This second stage cariostatic effect must be due to involvement of water from northwestern Ohio either in cooking or in drinking, or both.

From this experiment, one is unable to specify which element or combination of elements are involved, but, as was shown earlier, the elements that distinguish the waters of northwestern Ohio from the waters in the seven largest cities of Ohio and the high caries states were strontium, molybdenum, boron, and lithium, each of which has been shown to have an effect on caries by other experiments.

More specific knowledge as to the quantity of each element required to obtain the cariostatic effect will be available when we have evaluated the chemical analyses of the various diets that were performed for us by Professor J. Benton Jones, Jr., in the Department of Agronomy of the Ohio State University.

The fact that these four elements are simultaneously present in the water and the diet and that the concentrations of strontium, molybdenum, lithium, and fluoride are further increased by cooking suggests further research.

This summer we hope to repeat the rat-feeding experiment, but we will use aliquots of green beans derived from a single batch of seed, grown for us on different soils, as some of the data show considerable differences in the trace-element composition of green beans collected from different regions.

For the geographic pathologist, the question of whether or not the geologic environment in which one lives can produce a measurable impact on health is an intriguing one.

In collaboration with Allaway and others (1968) and Kubota and others (1968) at the U.S. Plant, Soil, and Nutrition Laboratory, Agricultural Research Service, and Soil Conservation Service, Ithaca, New York, an experiment was designed to determine whether or not there were geographic differences in trace-element concentrations in blood associated with geographic differences in trace-element concentrations in plants. Several elements were studied, but we shall discuss only the trace-element selenium as an example. Pilot tubes of whole blood were obtained from blood banks making collections in 19 different cities in the United States. The donors were adult male residents of their specific area. Selenium was determined by the fluorometric method of Allaway and Cary (1964). The mean for the 210 samples was 20.6 μg/100 ml, and the range was from 10 to 34 μg/100 ml. There were definite geographic variations in the concentrations of selenium in blood from different collection sites. Blood samples from Rapid City, South Dakota (25.6 μg/100 ml), Cheyenne, Wyoming (23.4 μg/100 ml), and Fargo, North Dakota (21.7 μg/100 ml), all of which are located in or near to areas where high-selenium vegetation has been found, contained significantly (0.05) more selenium than did blood samples from Muncie, Indiana (15.8 μg/100 ml), and Lima, Ohio (15.7 μg/100 ml), located in an area where plants are generally low in selenium.

Lima, Ohio, is in the caries-resistant, low-selenium area, and it is interesting that Hadjimarkos and Bonhorst (1958) have associated increased caries susceptibility with high-selenium intake.

The blood-selenium experiment emphasizes our need for much more geographic data on water, ready-to-eat food, body fluids, and tissues. Residents of Rapid City, South Dakota, and Cheyenne, Wyoming, were shown to have the highest blood-selenium concentration of the 19 collection sites studied. In a report by Greene and others (1965) of the Congenital Anomalies Section, Epidemiology Branch, U.S. Public Health Service, the states of Wyoming and South Dakota have the highest reported rates per 100,000 live births for reduction deformities and polydactylia. As the rates are calculated from birth certificates, their accuracy and completeness are somewhat questionable, but, assuming that these apparent geographic differences are real, additional blood studies from persons in many geographic areas should be made and compared with statistics on the distributions of congenital malformations in animals and man.

Ohio offers another area for study that relates to the geographic distribution of the number of deaths from arteriosclerotic heart disease (ISC 420). Through George Peoples, Jr., Chief Statistician for the Ohio Division of Vital Statistics, we obtained the number of deaths and the death rates per 100,000 population among white male residents, aged 45–64, in selected counties for each year since 1949. Several interesting patterns were noted. For example, some counties had rates almost equal to the average state rate for each year, whereas others had much lower rates during the earlier years, but then abruptly changed, maintaining significantly higher rates during later years. If these death rates reflect true differences, it would seem that an intrastate study might reveal the possible reasons for the changes in rates. Were the changes related to the introduction of a new water supply or the establishment of a new industry or refinery or change in population? These factors should be investigated as possibly they may be causally related to an important cause of mortality.

EPILOGUE

We have tried to point out what can be accomplished by investigators from different disciplines who work together on particular health problems and also what should be done in a collaborative attack on some of the most important diseases of our time.

Projects should be designed to include and to develop as much basic background information as possible. The project on the relationship between dental caries and environmental factors is an example. It deals with the relationship of soil, water, and foodstuffs and the state of the teeth. But it was planned that it could establish valuable basic information about the environmental impact on the food chain, which may be of use to workers investigating other areas of health and disease.

The study of trace element nutrition and human health is as yet in its infancy, and it would seem likely that further associations will be established in the future between trace-element intake and human disease. We believe that a pooling of resources between geochemists acquainted with our physical environment, and medical, dental, and animal investigators may provide a profitable way to approach geographic pathology.

REFERENCES CITED

Allaway, W. H., and Cary, E. E., 1964, Determination of submicrogram amounts of selenium in biological materials: Anal. Chemistry, v. 36, p. 1359–1362.

Allaway, W. H., Kubota, J., Losee, F., and Roth, Margaret, 1968, Selenium, molybdenum and vanadium in human blood: Arch. Environ. Health, v. 16, p. 342–348.

Büttner, W., 1963, Action of trace elements on the metabolism of fluoride: Jour. Dent. Research, v. 42, p. 453.

Greene, J. C., Vermillion, J. R., and Hay, Sylvia, 1965, Utilization of birth certificates in epidemiologic studies of cleft lip and palate: Cleft Palate Jour., v. 2, p. 141–156.

Hadjimarkos, D. M., and Bonhorst, C. W., 1958, The trace element selenium and its influence on dental caries susceptibility: Jour. Pediatrics, v. 52, p. 274–278.

Hubbell, R. B., Mendel, L. B., and Wakeman, A. J., 1937, A new salt mixture for use in experimental diets: Jour. Nutrition, v. 14, p. 273–277.

Kubota, J., Lazar, V. A., and Losee, F., 1968, Copper, zinc, cadmium, and lead in human blood from 19 locations in the United States: Arch. Environ. Health, v. 16, p. 788–793.

Losee, F. L., and Adkins, B. L., 1968, Anticariogenic effect of minerals in food and water: Nature, v. 219, p. 630–631.

——, 1969, A study of the mineral environment of caries-resistant navy recruits: Caries Research, v. 3, p. 23–31.

Martin, D. J., 1951, Fluorine content of vegetables cooked in fluorine-containing waters: Jour. Dent. Research, v. 30, p. 676–683.

Shaw, J. H., and Griffiths, D., 1961, Developmental and post-developmental influences on incidence of experimental dental caries resulting from dietary supplementation by various elements: Arch. Oral Biology, v. 5, p. 301–313.

Skougstad, M. W., and Horr, C. A., 1963, Occurrence and distribution of strontium in natural water: U.S. Geological Survey Water-Supply Paper 1496-D, 97 p.

Smith, H. V., Smith, M. C., and Vacich, M., 1945, Fluorine in milk, plant foods and foods cooked in fluorine-containing water: Arizona Agric. Exp. Station, Mimeo. Rept., 37 p.

MANUSCRIPT RECEIVED BY THE SOCIETY OCTOBER 8, 1969

Printed in U.S.A.

THE GEOLOGICAL SOCIETY OF AMERICA, INC.
MEMOIR 123, 1971

Medical Geography and Its Geologic Substrate

R. W. ARMSTRONG
Department of Geography and School of Public Health
University of Hawaii, Honolulu, Hawaii

ABSTRACT

Geographical associations between geochemical variables and the occurrence of certain human diseases have been pointed out for several decades. Associations between distributions of elements in soil and water and distributions of mortality from cardiovascular disease, multiple sclerosis, and certain cancers, have been particularly strong and persistent over time. Results of analytical studies have so far been inconclusive. Other relationships, apart from the geochemical, may be more important. A comparison of county distributions in Norway, Sweden and the Netherlands for a number of environmental variables, and prevalence of multiple sclerosis is presented. It is suggested that an ecological approach to the possible geochemical association with disease is critical, and that more attention could be given to that side of the relationship concerned with man.

CONTENTS

INTRODUCTION

Medical geography forms a distinct field of study largely because of the existence of intriguing geographical associations between human health and disease and certain environmental variables, such as the geochemical. It combines branches of medical science and geography to examine the spatial variations of both the environment in relation to health, and health as a characteristic of the environment.

From an ecological perspective, human health can be viewed as a state of adjustment by the organism to its own internal bodily environment and to the external environment. The external environment is taken here to comprise not only the physical and biological realms, but the cultural and conceptual realms of man as well. In this context, disease may be taken as some degree of failure by the organism to adjust to challenges posed by the internal or external environments, or both (Sargent and Barr, 1965). Among populations, the degree of failure to adjust to some of these challenges is reflected crudely by morbidity and mortality indices which exhibit variations in rate through space and time.

The spatial distribution of morbidity and mortality is nonrandom, irrespective of the scale employed (Sargent, 1964), and many of the variables in the external environment which have been hypothesized as challenges to health also follow nonrandom spatial distributions. One objective in medical geography is to isolate and identify those variables in the external environment, which co-vary geographically with disease indices, and which could conceivably be concerned in disease causation. A second objective is to help in indicating the relative importance of environmental variables external to the organism and those which are internal or constitutional.

While medical geography is, in theory, concerned with all kinds of variables in the external environment, it has given particular attention to those relating to geology. This is in part perhaps a reflection of the traditional emphasis given by physical geography to geomorphology, and the fact that geological distributions were among the first elements of the physical environment to be precisely mapped. No doubt a more pertinent reason is that the association with geochemical variables has an acceptable mechanism of connection through the food chain.

GEOGRAPHICAL ASSOCIATIONS

With few exceptions the medico-geographical studies concerned with a geochemical relationship have dealt with chronic diseases of little-known etiology. Cancer, cardiovascular disease, and multiple sclerosis have attracted most attention because they exhibit pronounced regional variations in rate, and because there is strong evidence that factors in the external environment are implicated in their etiology. In the case of cancer, it has been estimated that about two-thirds of tumors in man in the western countries are due to environmental factors, while racial and genetic factors appear to be of secondary importance (Higginson, 1967). More specifically, the dietary factor has been suggested as important by a number of studies. This, in turn, has led to hypotheses that the chemical nature of rock and soil may be associated with the incidence of cancer, as well as certain other diseases, through the mechanism of the food chain.

Until the advent of spectrographic and other techniques for handling mass samplings for chemical analyses, most of the searching for relationships between geology and disease had utilized comparative mapping surveys. The persistence of many of the associations that these surveys have suggested is striking. Haviland (1892) first pointed out a geographical association between cancer mortality and geologic features in England in 1868, and since that time comparable relations at similar levels of generalization have continued to be reported at various times and in different localities (Armstrong, 1962). A recent mapping survey by Takahashi (1967) compared the distribution of mortality in 1964 for cerebrovascular disease and arteriosclerotic heart disease within 18 European countries and Japan. Geographic associations between the distributions of mortality and distributions of geologic and climatic features, character of water, and socio-economic factors were suggested.

During the last decade there have been a number of studies of the geochemical-disease relationship which are more analytical. These have shown an increasing concern with such questions of research design as comparable levels of measurement, adequate comparative controls, probability sampling, and standard forms of analysis. Most of these recent studies have confined their attention to a specific disease in a selected community. In the case of cancer, for example, Stocks and Davies (1960) analyzed soil elements in samples from 300 home gardens in Wales and England and found high levels of chromium, zinc, cobalt, and organic matter associated with mortality from cancer of the stomach. However, in rural Iceland, among a small group of matched cases and controls, Armstrong (1964; 1967) found no consistent relationships between mortality from stomach cancer and the trace element contents of samples of pasture grass, milk, and drinking water. In the Transkei of South Africa, Marais and Drewes (1962) found a significant geographical association between high frequencies of cancer of the oesophagus and areas of declining soil fertility and severe soil erosion. In short, conclusions from studies of cancers of different types show partial, but no conclusive, overlap, and there are some contradictions.

The variations in cardiovascular disease mortality, on the other hand, show a more consistent geographical association with a particular geochemical variable, namely the relative hardness of drinking water. Studies in the United States, England and Wales, Sweden, and Japan, indicate that communities with soft-water supplies have higher mortality rates from the disease. The most conclusive works are those of Biörck and others (1965) in Sweden, Schroeder (1966) in the United States, and Crawford and others (1968) in England and Wales; all of which indicate that levels of calcium are associated with the variance of the cardiovascular death rate. No evidence of confounding by other known environmental factors was revealed in the British study, and the general conclusion is that there appears to be a real association between some unknown factor (or factors) in water and cardiovascular disease. Considerable research into the possible geochemical relationships of a number of diseases also has been reported from the Soviet Union (Gelyakov and others, 1966).

ASSESSMENT

Because of differences in subject, time period, and research designs, it is difficult to compare the medico-geographical studies on geochemical relations with disease which have been reported thus far. Evidence suggesting that associations exist continues to accumulate, but the geographical and ecological significance remains obscure. There are several likely reasons for these inconclusive results, and a brief review of some of them may serve to point up the need for a more rigorous and interdisciplinary approach.

First, there is a need to formulate more specific and meaningful hypotheses for testing in medical geography. This can best come through basic research into the general associations now reported which should identify clear propositions and implicate particular mechanisms for subsequent testing. Until a close reciprocity is established between work in basic research and in the field, we cannot expect much progress in refining and testing the current body of knowledge based on vague associations.

Secondly, there have been obvious difficulties in securing reliable and comparable data for medico-geographical studies. Survey sampling techniques, which have been successfully used in medical research in a range of situations, including Africa, offer a possible solution to the problem of obtaining reliable data specifically designed to test geochemical associations in contrasting external environments. Geographical associations must be scrutinized to ensure that they are real and not just due to some statistical artifact (Sauer and Enterline, 1959), and to ensure that the association holds at different geographical scales and levels of measurement (Kurtzke, 1967a). Disease distributions may also change radically over short time periods, and significant associations with environmental variables in one decade may not be valid for the next (Momiyama-Sakamoto, 1966).

OTHER POSSIBLE RELATIONSHIPS

Geochemical associations may be only coincidental, or, at best, only partially correlated with disease indices, and other environmental or human constitutional factors may be more important. To some extent this question can be dealt with by a critical assessment of the relationships suggested for all kinds of variables in the external environment from information and theory concerning mechanisms of disease causation. Sometimes, established risk factors for particular diseases can be applied to see whether or not they explain geographical variations in the disease frequency. For example, in the United States, per capita cigarette sales by state have been shown to correlate strongly with mortality rates from coronary heart disease (Friedman, 1967).

Exhaustive reviews of world-wide evidence for environmental relationships have recently been written for cancer of the intestine (Wynder and Shigematsu, 1967) and sarcoidosis (Siltzbach, 1967). In the case of intestinal cancer, dietary factors were reaffirmed as the most likely environmental relationship, although this is by no means proven. In the case of sarcoidosis, the recent survey suggests that certain ethnic susceptibilities may explain the intriguing geographic distribution of the disease, but no evidence was found to support the conclusion of an earlier paper (Gentry and others, 1955) that there was a relationship between sarcoidosis and soil type in the southern United States.

The prevalence of multiple sclerosis has a peculiar geographic distribution as evidenced by detailed data for the Scandinavian countries, Switzerland, and the Netherlands (Kurtzke, 1967a). A variety of environmental variables appear to co-vary geographically with the disease and these have been used in support of a number of tentative hypotheses for its cause. Certain geochemical associations have been suggested. One by Warren (1963) pointed out that communities in Sweden and Norway with a low prevalence of the disease were living in geological areas which had rock with low lead content. These associations lend weight to a hypothesis that the disease is caused by some factor in the diet. Other studies tend to support a hypothesis that multiple sclerosis is an infectious disease of childhood with a long period of latency before becoming evident later in life. Kurtzke (1968) found modest statistical associations between the distribution of multiple sclerosis and measles, mumps, and scarlatina in Norway, Denmark, and Switzerland. In a unique study of multiple sclerosis among immigrants to Israel, Liebowitz and others (1967) correlated twenty environmental variables with the prevalence rates of the disease according to country of origin of the immigrants. Twenty-two countries in Europe, the Middle East and North Africa were used in the analysis. The highest rates of the disease were among those immigrants from Northwestern and Central Europe, and there were strong positive correlations with socioeconomic variables which were indicative of high standards of living. It was postulated that these correlations were compatible with a hypothesis that multiple sclerosis may be a relatively frequent disease in communities where improved hygiene reduces or delays exposure of an individual to the disease-causing factors, whatever they may be.

It seemed of interest to see if the findings of Liebowitz and others (1967) could be recognized within particular countries on a much larger scale. For example, would the correlations found between the distribution of multiple sclerosis and environmental variables for European and neighboring countries recur if small units, such as counties, were the basis for comparison? A preliminary examination of the question was made for Norway, Sweden, and the Netherlands. The prevalence rates of multiple sclerosis, by county units, for the three countries have been reported by Kurtzke (1967b). They were correlated with a limited number of statistics for environmental variables, and the results indicated only slight support of the general conclusion that the disease is associated with higher standards of living in both urban and rural communities (Table 1). The correlation procedure ranked counties in each country in

TABLE 1. CORRELATION OF SELECTED ENVIRONMENTAL VARIABLES WITH PREVALENCE RATES OF MULTIPLE SCLEROSIS FOR A GROUP OF EUROPEAN COUNTRIES AND FOR COUNTIES OF NORWAY, SWEDEN, AND THE NETHERLANDS

Variable	Multiple Sclerosis Prevalence: Period and Area			
	1961 22 European and Neighboring Countries r*	1935–48 Norway (Rural cases) r_s†	1925–34 Sweden r_s†	1946–55 Netherlands r_s†
Sunshine (mean annual hours)	−.88‡			
Mean annual temperature	−.70‡	.09	−.10	
Mean January temperature		−.18	−.13	
Mean July temperature		.57	.04	
Percent area young sea clay				.67
Percent area sand				−.42
Percent area peat				.04
Population density per km^2		.22	.02	.04
Proportion population 5–14 years	−.81‡		−.42	−.55
Infant mortality (per 1000 live births)	−.69‡	−.23	.00	−.70
Steel consumption (tons per 1000 population)	.85‡			
Electricity consumption (per capita)	.44§	.35		
Motor vehicle density (vehicles per 1000 population)	.44§	.78	.59	.37
Cultivated land (hectares per capita)	−.39	.40	.27	−.03
Wheat yields (Kg per hectare of cultivated land)	.63‡		.12	
Grain production (Kg per hectare of cultivated land)		.58		
Fertilizer consumption (Kg per hectare of cultivated land)	.66‡	.26		

*Correlation coefficients for 22 countries of Europe, the Middle East and North Africa from Liebowitz and others, 1967

†Spearman rank correlation coefficients, r_s, were used in this preliminary survey of county distributions within Norway, Sweden and the Netherlands instead of product moment correlation coefficients, r, because the data for multiple sclerosis and environmental variables were at different time periods. It is expectable that changes over time in the rank of environmental variables would be less significant than changes over time in magnitude. Prevalence rates for multiple sclerosis by county for Norway, Sweden and the Netherlands are reported by Kurtzke (1967b) from the work of others. Data were drawn from hospital clinical records. Temperature data for Norway and Sweden were taken from: International Agro-Climatological Series, Washington, D.C., Am. Inst. Crop Ecology, 1950; and data for other variables for Norway, Sweden, and the Netherlands from official statistical yearbooks, 1960–62.

‡Significant at the 1 per cent level

§ Significant at the 5 per cent level

terms of their multiple sclerosis prevalence rate, then, according to the values reported for the environmental variables. The county ranking for each environmental variable was compared against the ranking for multiple sclerosis, and the degree of correlation was measured using Spearman's rank correlation coefficient.

The rank correlation values were much weaker in this detailed comparison, which uses data from entire populations rather than from samples. These results may be due in part to a difference in time period between data for the disease and for the environmental variables, or to the comparative uniformity and lack of contrast in the socio-economic conditions of these particular countries.

AN ECOLOGICAL APPROACH

At the present time, none of the geographical associations that have been demonstrated between various environmental variables and multiple sclerosis, cancer, and cardiovascular disease can be shown to have any significance in the etiology of these diseases. The relative importance of geochemical variables *versus* other environmental factors is still a guess. The only point of general agreement is that these chronic diseases have a complex, multifactorial etiology which makes it important to examine any particular set of variables in their ecological context.

Figure 1 places the food chain—the logical mechanism for geochemical implications with disease—in a general ecological frame. Human ingestion of rock-derived elements, that may or may not influence the susceptibility or resistance of the body to disease, is influenced by a host of physical, biological, and cultural factors ranging from climate, through plant and animal metabolism to cooking practices. Further, there are the individual characteristics of man himself, particularly his physiology, his selectivity, and his capacity to adapt to environmental challenges. Time and space are significant in this system because events move at different rates and rhythms and have

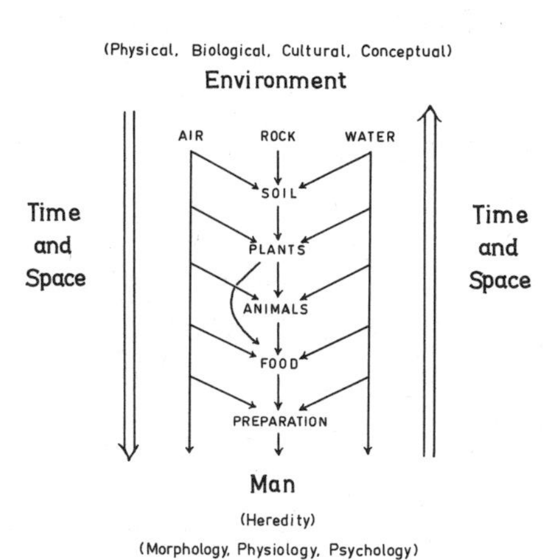

Figure 1. Generalized model of the transport of geochemical elements through the food chain as one component of a human ecological system. Single lines sketch the food chain; double lines, other pathways of interaction. Heredity, morphology, physiology, and psychology represent the origin, form, function, and behavior of man. We do not know if psychology, or the conceptual environment, that is, how we view things, has any direct role in the geochemical relationship, but a human ecological system is not complete without them.

wide-ranging origins. A particular person's diet, for example, will usually originate from a number of localities both near to and far from his normal place of residence. To explore the intricacies of this ecological system would require the combined efforts of scientists drawn from several basic and applied fields, each employing coordinated research strategies which would allow subsequent integration of the separate studies to provide a better understanding of the whole system. An interdisciplinary approach seems the only feasible way of probing further into the question of geochemical implications with disease, of evaluating their relative importance to other environmental variables, and of proving or refuting the validity of the associations which appear suggestive of vital clues to the etiology of major chronic diseases.

CONCLUSIONS

The medical geographer's task in this coordinated research is to analyze and describe communities and their environments in terms which are specific to the geochemical relationship with health and disease. For instance, the validity of relationships in particular localities may depend on the proportion of locally grown food in the diet. Food origin surveys, evaluating differences from place to place in sources of foodstuffs, home garden practices, and water supply, could identify communities where geochemical studies would have most relevance (Armstrong, 1970). There are also considerable variations in dietary and culinary practices from place to place which may bear on the validity of particular geochemical and disease relationships. The distributions of certain physical and biological variables, such as climatic elements and the associations of plants and animals, may also be relevant.

Data collection methods and analytical approaches need improvement. A series of comparable maps dealing with specific topics could summarize the available geographic information on the geochemical relationship with disease. Such a series could cover the range of the relationship from geology, climate, and soils, through vegetation, agriculture, and food handling, to human health and disease. The methods of testing geographical associations between environmental variables and disease must be refined and extended, and linked more closely to theoretical mechanisms of cause and effect proposed by basic research.

REFERENCES CITED

Armstrong, R. W., 1962, Cancer and soil: Review and counsel: Professional Geographer, v. 14, p. 7–13.

——, 1964, Environmental factors involved in studying the relationship between soil elements and disease: Am. Jour. Public Health, v. 54, p. 1536–1544.

——, 1967, Milk and stomach cancer in Iceland: Acta Agric. Scandinavia, v. 17, p. 30–32.

——, 1970, Local food sources in a southern Illinois population: Geographical Analysis, v. 2, p. 149–164.

Biörck, G., Boström, H., and Widström, A., 1965, On the relationship between water hardness and death rates in cardiovascular diseases: Acta Med. Scandinavia, v. 178, p. 239–252.

Crawford, M. D., Gardner, M. J., and Morris, J. N., 1968, Mortality and hardness of local water supplies: Lancet, 7547, v. 1, p. 827–831.

Friedman, G. D., 1967, Cigarette smoking and geographic variation in coronary heart disease mortality in the United States: Jour. Chronic Diseases, v. 20, p. 769–779.

Gelyakov, T. M., Voronov, A. G., Nefedova, V. B., and Samoylova, G. S., 1966, The present state of medical geography in the U.S.S.R.: Vestnik Moskovskogo Universiteta, seriya geografiya, no. 4, p. 28–33, (English translation: 1967, Soviet Geography, v. 8, p. 228–234).

Gentry, J. T., Nitowsky, H. M., and Michael, M., 1955, Studies on the epidemiology of sarcoidosis in the United States: The relationship to soil areas and to urban-rural residence: Jour. Clin. Invest., v. 34, p. 1839–1856.

Haviland, A., 1892, Geographical distribution of disease in Great Britain, 2nd Edition: London, Sonnenschein.

Higginson, J., 1967, The role of geographical pathology in cancer research: Schweizer Med. Wochenschrift, v. 97, p. 565–568.

Kurtzke, J. F., 1967a, On the fine structure of the distribution of multiple sclerosis: Acta Neurology Scandinavia, v. 43, p. 257–282.

——, 1967b, Further considerations on the geographic distribution of multiple sclerosis: Acta Neurology Scandinavia, v. 43, p. 283–298.

——, 1968, Multiple sclerosis and infection from an epidemiologic aspect: Neurology, v. 18, p. 170–175.

Liebowitz, U., Sharon, D., and Alter, M., 1967, Geographical considerations in multiple sclerosis: Brain, v. 90, p. 871–886.

Marais, J. A. H., and Drewes, E. F. R., 1962, The relationship between solid geology and oesophageal cancer distribution in the Transkei: South Africa Geol. Survey Annals, v. 1, p. 105–114.

Momiyama-Sakamoto, M., 1966, A study in methodology of medical geography: Acta Geol. et Geog. Universitatis Comenianae. Geog., nr. 6, p. 213–223.

Sargent, F., 1964, The environment and human health: Arid Zone Research, v. 24, p. 19–32.

Sargent, F., and Barr, D. M., 1965, Health and fitness of the ecosystem: The Environment and Man, Hartford, Travelers Research Center, p. 28–46.

Sauer, H. I., and Enterline, P. E., 1959, Are geographic variations in death rates for the cardiovascular diseases real?: Jour. Chronic Diseases, v. 10, p. 513–524.

Schroeder, H. A., 1966, Municipal drinking water and cardiovascular death rates: Am. Med. Assoc. Jour., v. 195, p. 81–85.

Siltzbach, L. E., 1967, Geographic aspects of sarcoidosis: New York Acad. Sci. Trans., v. 29, p. 364–374.

Stocks, P., and Davies, R. I., 1960, Epidemiological evidence from chemical and spectrographic analyses that soil is concerned in the causation of cancer: British Jour. Cancer, v. 14, p. 8–22.

Takahashi, E., 1967, Geographic distribution of mortality rate from cerebrovascular disease in European countries: Tohoku Jour. Experim. Medicine, v. 92, p. 345–378.

Warren, H. V., 1963, Trace elements and epidemiology: Jour. College Gen. Practitioners, v. 6, p. 517–531.

Wynder, E. L., and Shigematsu, T., 1967, Environmental factors of cancer of the colon and rectum: Cancer, v. 20, p. 1520–1561.

MANUSCRIPT RECEIVED BY THE SOCIETY JANUARY 5, 1970

Printed in U.S.A.

THE GEOLOGICAL SOCIETY OF AMERICA, INC.
MEMOIR 123, 1971

Discussion

James Kmet, M. D. *Epidemiology Section, International Association for Research on Cancer, Lyon, France*

As has been discussed, cancer is an important disease, some forms of which may be causally connected to the geochemical environment. I should like to mention a study in progress that may prove to be pertinent to the principal subject under discussion.

The International Agency for Research on Cancer and the Institute of Public Health Research, Teheran, is conducting a collaborative research project in the Caspian Province of Mazandaran, where there appears to be a fairly sharp borderline between a high oesophageal-cancer incidence area in the east and a low incidence region in the west. Although these findings remain to be confirmed, they have stimulated a broad environmental study of the differences in the physical, biotic, and cultural characteristics between the low and high incidence areas.

Takahisa Hanya *Department of Chemistry, Tokyo Metropolitan University, Tokyo, Japan*

I would like to comment upon two interesting chronic diseases of my country, one of which is definitely causally related to geochemical environment, the other of which is, as yet, of unknown etiology.

The first is *Itaiitai* disease. The Japanese word, *Itaiitai,* means pain or painful. The patient of this disease is said to have fragile bones and to be in great pain. Patients of this disease number 223 at the present time. The disease occurs only in a very limited area, where the cadmium content of soils in rice fields is unusually high, averaging about 3.0 ppm. The rice, which is harvested from this area, is an important food for the native Japanese and contains an average of 2 ppm cadmium. Recently it has been concluded that the high concentration of cadmium in food is one of the important causes of this disease. The cadmium in the soils of this area has built up over a long period from the waste water of the Kamioka mine.

The second condition is Kashin-Beck disease. This chronic bone disease was first found in Siberia, USSR, but we have found many cases of the disease in various places in Japan. Its cause is not yet clear. The symptoms of the disease are, in general, so slight that many who contract it do not consult a physician. However, the condition can be detected by X-ray inspection. Research conducted by Professor Takizawa suggests that some kinds of organic acids contained in drinking water cause this disease. These substances are now being separated and identified in my laboratory, but the research is not yet concluded. Substances causing Kashin-Beck disease are found in both natural humic waters and polluted waters containing municipal sewage.

P. W. Hall, M.D. *Case Western Reserve University, Cleveland Metropolitan General Hospital, Cleveland, Ohio*

I should like to describe briefly a most interesting disease, endemic Balkan nephropathy, that may be caused by an unfavorable geochemical environment.

Twelve years ago clinicians in Bulgaria, Romania, and Yugoslavia

Figure 1. Geographic locations (dotted lines) of endemic renal disease in river valley areas in Yugoslavia and Romania (*from* Ann. Conf. on the Kidney Proc., Natl. Kidney Foundation, Fig. 96, p. 313).

described an endemic renal disease localized geographically in river valley areas in Yugoslavia and Romania and in the foothills of the Balkans in Bulgaria (Fig. 1). This disease is characterized by the insidious onset of uremia. Fatigue and backache are the only symptoms. Anemia, azotemia, acidosis, and mild proteinuria are the major medical signs. Aside from the kidneys, no other organs appear to be involved. Hypertension is unusual. The disease is common in families, but is not familial. It is predominantly a tubular-interstitial fibrosis without inflammation. The disease usually is detected in the third or fourth decade of life. From clinical symptoms to death is usually 3 to 5 yrs. At autopsy the kidneys weigh 20 to 40 grams (normal 125 to 150 grams).

By studying the electrophoretic distribution and molecular size of the urinary proteins, it is evident that there is an increased excretion of low molecular-weight proteins, strongly suggesting tubular damage. Using these techniques, we have shown that: (1) Approximately 50 percent of the individuals living in certain villages along the Sava and Drina Rivers of Yugoslavia have evidence of this damage. (2) In many, the evidence is seasonal, being positive in spring and fall, but negative in midwinter. (3) Only farm field workers have the disease. Members of the same family who eat the same food and drink the same house water, but who do not work on the farms, are not affected.

The two major differences apparent between the farmer and nonfarmer are: (a) farmers are in close physical contact with the land, obviously; (b) their field water supply is different from that of the home.

The disease resembles that which results from certain types of chronic metal poisoning—cadmium and uranium.

Epidemiologic studies of one entire village revealed that: (1) Poor farmers working poor soil that is flooded two or three times each season, suffer most with the disease. (2) Sex distribution is equal, but morbidity and mortality rates are higher among females. (3) No evidence of this disease exists in any villager who does not actually farm the land. (4) Not all farmers are involved. (5) In Yugoslavia, prevalence of the disease decreases as the altitude increases, reaching zero at approximately 100 ft above the river level.

Our conclusions, so far, are these: (1) This condition does not follow the epidemiology of any known bacterial infection of the human kidney. No virus has yet been incriminated in the etiology of chronic renal failure in human beings. (2) The lesion in the kidney resembles that seen in damage due to chronic low-level trace-element poisoning. (3) Studies of drinking water, soil, plant, and kidney tissues for trace-element content, to date, have been unrewarding.

MANUSCRIPTS RECEIVED BY THE SOCIETY JULY 1, 1970

Printed in U.S.A.

Index